For the Rising Math Olympians

The Ultimate Handbook for Winning Math Competitions in Elementary and Middle School

Jesse Doan

ISBN-13: 978-1536991079
ISBN-10: 1536991074

For information or questions, please contact the author, Jesse Doan, at risingmatholympians@gmail.com.

Table of Contents

Introduction

In fifth grade, I started training for MATHCOUNTS®* and the American Mathematics Competitions, the premier middle school math contests. In order to prepare for these contests, I completed over ten math workbooks. From my personal experience, I found a lack of math competition resources for beginners. As a result, I was compelled to write this book for elementary and middle school students to bridge school curriculum math and competition math.

For the past three years, I have served as an assistant coach for my former middle school math team. From this coaching role, I learned that many beginning mathletes find competition problems to be too challenging or impossible for them to solve. Consequently, I structured this book to cover the basic concepts that mathletes need to master before approaching math competition problems.

This book covers 20 math topics in Number Theory, Algebra, Counting & Probability, and Geometry that are frequently tested in math competitions. Each chapter contains concepts with detailed explanations, examples with step-by-step solutions, and review problems to reinforce the students' understanding. The most challenging problems are denoted with asterisks (*).

It is highly recommended that the students working through this book are proficient in addition, subtraction, multiplication and division. Furthermore, knowledge in pre-algebra would be helpful. Some chapters build on concepts from previous chapters, therefore, I suggest that students work from the beginning to end.

I hope this book will serve as a valuable resource for young ambitious students who desire to excel in math competitions. This book may also serve as an accessible tool for math team coaches who wish to teach their students advanced problem solving skills in preparation for elementary and middle school math competitions.

Jesse Doan
August 2016

*MATHCOUNTS® is a trademark registered by the MATHCOUNTS Foundation.

Acknowledgements

First of all, I would like to thank the Lord for all of His wisdoms, favors, and blessings. I would also like to thank my parents and my siblings for their continuous support and encouragement. Lastly, I would like to thank my math coaches for inspiring me to help other young mathletes.

Math Vocabulary and Order of Operations

Math Terms

The terms listed below will appear in this book and in math competitions.

The Basic Terms:

Positive Numbers:	Any number greater than 0. {1, 2, 4.5, … }
Zero or 0:	The only real number that is neither positive nor negative.
Negative Numbers:	Any number less than 0. {−1, −3, −4.5, … }

Natural Numbers or $\mathbb{N}$ consist of all the positive numbers excluding fractions and decimals. They are also known as counting numbers.
$\mathbb{N}$ | {1, 2, 3, 4, 5, … }

Whole Numbers consist of 0 and the Natural Numbers.
Whole Numbers | {0, 1, 2, 3, 4, 5, … }

Integers or $\mathbb{Z}$ consist of 0, all positive and negative numbers excluding fractions and decimals.
$\mathbb{Z}$ | { … −5, −4, −3, −2, −1, 0, 1, 2, 3, 4, 5, … }

Rational Numbers or $\mathbb{Q}$ consist of all the numbers that can be expressed as a ratio of two integers. For example, a simple fraction is a rational number. In addition, integers are rational numbers because they can be expressed as a ratio of the integer and 1.
{i.e. ½, 5, $-\frac{13}{8}$}

Irrational Numbers or $\mathbb{I}$ consist of all the numbers that cannot be expressed as a ratio of two integers.
{i.e. π, $\sqrt{2}$}

Real Numbers or $\mathbb{R}$ are the combination of the rational and irrational numbers.

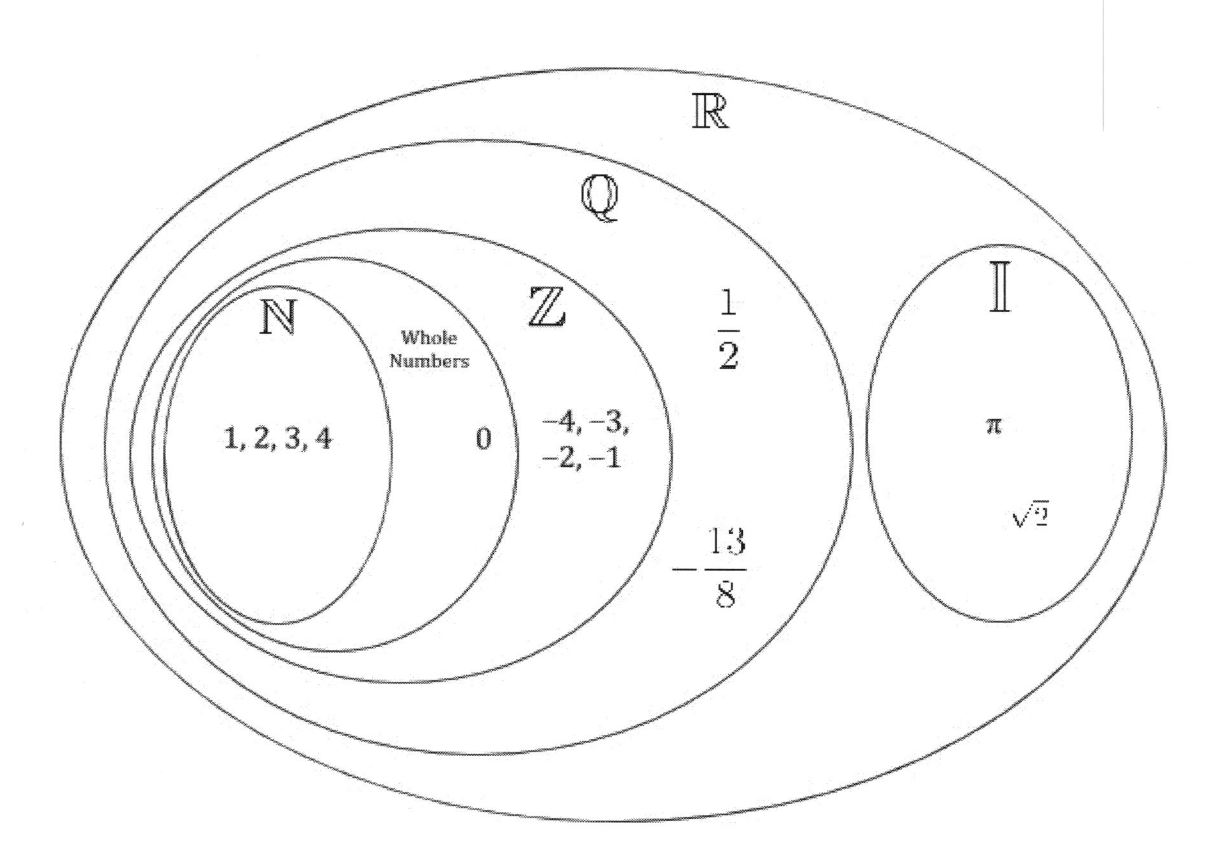

The Four Operations (+, –, ×, ÷)

Addition (+)

$1 + 2 = 3$

1 = augend; 2 = addend; 3 = **sum**

Subtraction (–)

$5 - 4 = 1$

5 = minuend; 4 = subtrahend; 1 = **difference**

Multiplication (× or ∗)

$2 \times 4 = 8$

2 = multiplicand; 4 = multiplier; 8 = **product**

Division (÷ or /)

$7 \div 2 = 3$ R 1

7 = dividend; 2 = divisor; 3 = **quotient**; 1 = **remainder**

(Tip: Bolded math vocabulary words should be memorized.)

Chapter 1

Order of Operations (also known as PEMDAS)

The Order of Operations consist of 4 steps:

Step 1) **P**arentheses: ( )
Step 2) **E**xponents (For example, $5^2 = 5 \times 5 = 25$ in which the exponent, 2, indicates how many times the number is multiplied.)
Step 3) (From left to right)
- **M**ultiplication: (×)
- **D**ivision: (÷)

Step 4) (From left to right)
- **A**ddition: (+)
- **S**ubtraction: (–)

Counting Numbers

When counting the number of integers between two numbers *a* and *b*, it can either be inclusive meaning the numbers *a* and *b* are included in the count or not inclusive.

The number of integers between *a* and *b*, inclusive, is $(b - a) + 1$.
The number of integers between *a* and *b* is $(b - a) - 1$.

For example, the number of integers between 1 and 10, inclusive, is 10. Using the formula: $(b - a) + 1 = (10 - 1) + 1 = 9 + 1 = 10$.
Another example, the number of integers between 1 and 10 is 8. Using the formula: $(b - a) - 1 = (10 - 1) - 1 = 9 - 1 = 8$.

Example 1: Solve: $2 \times (7 - 3) + 6 \div 2$

Solution:

1) Parentheses:	$2 \times (4) + 6 \div 2 = 2 \times 4 + 6 \div 2$
2) Exponents:	$2 \times 4 + 6 \div 2$
3) Multiplication/Division from left to right:	$8 + 6 \div 2 = 8 + 3$
4) Addition/Subtraction from left to right:	11

Answer: 11

Example 2: Solve: (17 – 5) ÷ 4 × 3 + 5

Solution:

1) Parentheses: (12) ÷ 4 × 3 + 5 = 12 ÷ 4 × 3 + 5
2) Exponents: 12 ÷ 4 × 3 + 5
3) Multiplication/Division from left to right: 3 × 3 + 5 = 9 + 5
4) Addition/Subtraction from left to right: 14

Answer: 14

Example 3: Jeremy read from page 24 to page 158, inclusive. How many pages did he read?

Solution: Using the formula $(b - a) + 1$,
$(158 - 24) + 1 = 134 + 1 = 135$

Answer: 135 pages

Example 4: Zachary counted all the integers between –45 and 233. How many integers did he count?

Solution: Using the formula $(b - a) - 1$,
$(233 - (-45)) - 1 = (233 + 45) - 1 = 278 - 1 = 277$

Answer: 277 integers

Example 5: Solve: $5 + 3 \times (3^3 - 25)^2$

Solution:

1) Parentheses: $5 + 3 \times (3\times3\times3 - 25)^2 = 5 + 3 \times (27 - 25)^2 = 5 + 3 \times (2)^2 = 5 + 3 \times 2^2$
2) Exponents: $5 + 3 \times 4$
3) Multiplication/Division from left to right: $5 + 12$
4) Addition/Subtraction from left to right: 17

Answer: 17

Review Problems

1. Solve: 1 – 2 + 3 – 4 + 5

   1-2 = -1+3 = 2-4 = -2+5 = 3

   1 - 2 + 3 - 4 + 5

   9 - 6 = 3

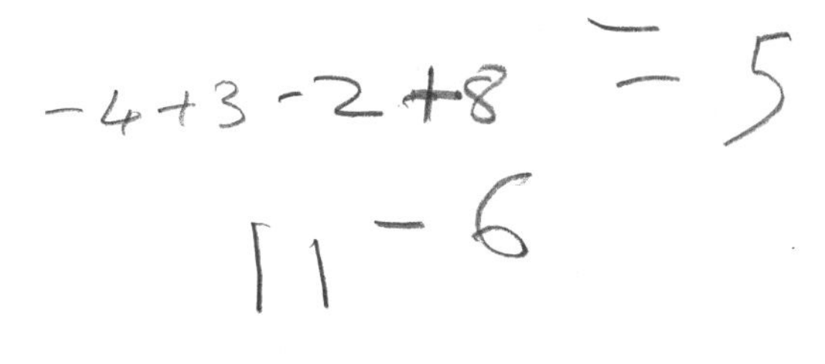

2. Solve: 1 + 2 × 2

   5

3. Solve: 1 – (2 + 3) – (4 + 5)

   1-5-9 = -13

4. Solve: (1 + 2) × 2

   3

   6

5. Solve: 12 ÷ 2 × 3

24

6. Solve: $0 + 2 - 4 \times 6 \div 8$

$0+2-24\div 8$

$0+2-3$

-1

7. Solve: $1 \times 2 + 3 \times 4 + 5 \times 6 + 7 \times 8$

$2+12+30+56$

100

8. Solve: $1 \times (2 + 3) \times 4 + 5 \times 6 + 7 \times 8$

$20+30+56$

106

9. How many integers are between -10 and 10?

$(b-a)+1$

19

10. How many natural numbers are between $-12{,}394$ and 932, inclusive?

$(b-a)+1$

$(932-1)+1$

932

11. Solve: 32 – (16 – (8 – (4 – (2 – 1))))
    Hint: Start with the parentheses that is nested the most.

12. Solve: $2^{((2+3) \times 4 - 18)}$

    $5 \times 4 - 18$

    $20 - 18 = 2$

    $2^2 = 4$

13. Solve: 6(2 + 4)
    Hint: 6(2 + 4) = 6 × (2 + 4).

    36

14. Solve: 20(50 + 4) – 1000

    20 (54) - 1000

    1080 - 1000

    80

15. Solve: 5(2 + 4 + 6 + 8 + 10)

16. Solve: $32 \div 16 \times 6 \div 3 \times 10 \div 5 + 18 \div 9 \times 12 \div 6 \times 8 \div 4$

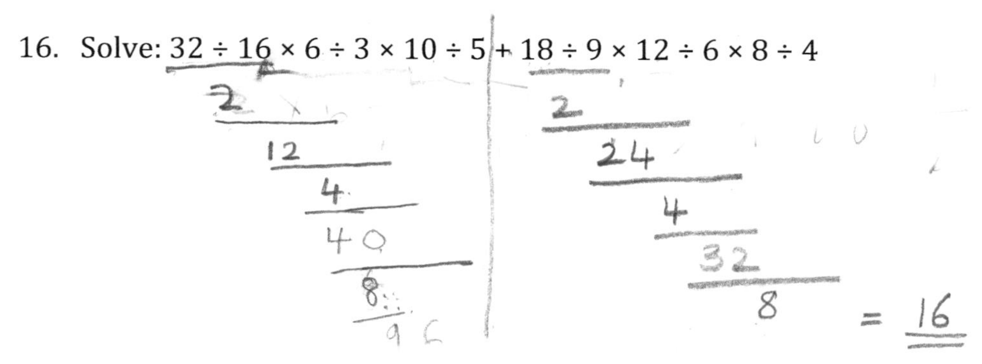

17. How many integers are between −34 and 89, inclusive?

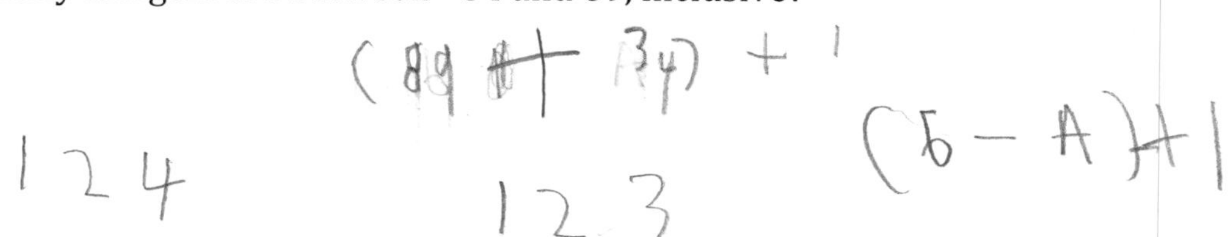

18. Jake has to read his book from page 186 to page 563, inclusive. If he wants to read the same number of pages for 14 days, how many pages should he read each day?

19. Solve: $16 \div (2 + 2)^2 + (19 - 2 \div 2) \div 6$

20. Solve: $720 \div 6 \div 5 \div 4 \div 3 \div 2 \div 1$

Divisibility Rules

What does "**divisible by**" mean?

"**Divisible by**" means when one integer is divided by another integer, the quotient is an integer. For example, given two integers a and b, if a is divisible by b, it means that when a is divided by b, the quotient is an integer and there is no remainder.
"a is **divisible by** b" also means "b is a **factor** of a" or "a is a **multiple** of b".

Divisibility Rules

A number is divisible by…

2: if the number is even or ends in 0, 2, 4, 6, or 8
3: if the sum of digits is divisible by 3
4: if the last two digits make a two-digit number that is divisible by 4
5: if the number ends in 0 or 5
6: if the number is divisible by 2 and 3
8: if the last three digits make a three-digit number that is divisible by 8
9: if the sum of digits is divisible by 9
10: if the number ends in 0
11: if the difference between the sum of the even-numbered digits and the sum of the odd-numbered digits is 0 or a multiple of 11

General Rule: The divisibility rule for an integer N, where N = p × q, is if the number is divisible by p and q. (For example: The divisibility rule for the number 6)
Therefore:
12: if the number is divisible by 3 and 4 (12 = 3 × 4)

Tip #1: A shortcut to find the remainder when a number is divided by 3 or 9 is to find the sum of the digits and divide it by 3 or 9.

Tip #2: A shortcut to find the remainder when a number is divided by 2, 5, or 10 is to find the last digit and divide it by 2, 5 or 10.

Tip #3: A shortcut to find the remainder when a number is divided by 4 is to find the two-digit number from the last two digits and divide it by 4.

Tip #4: A shortcut to find the remainder when a number is divided by 8 is to find the three-digit number from the last three digits and divide it by 8.

Example 1: Is 841 divisible by 3?

8 + 4 + 1 = 13
13 ÷ 3 = 4 R 1

Solution: We add the digits 8, 4, and 1 to test this. 8 + 4 + 1 = 13. 13 is not divisible by 3 so **841 is not divisible by 3**. Note: The remainder when 841 is divided by 3 is the same as when 13 is divided by 3 which is 1 (Tip #1!).

Answer: 841 is not divisible by 3

Example 2: Is 8,568 divisible by 9 and 8?

Solution: Let's add the digits to test if it's divisible by 9.
8 + 5 + 6 + 8 = 27. 27 is divisible by 9, so 8,568 is divisible by 9.

What about 8? Take the last three digits 568 and test for divisibility by 8.
568 ÷ 8 = 71 r 0

Answer: 8,568 is divisible by 8 and 9

Example 3: Which of these numbers {2, 3, 4, 5, 6, 8, 9, 10} is 180 divisible by?

Solution: 180 ends in 0 so it is divisible by 2, 5 and 10.
The sum of the digits is 1 + 8 + 0 = 9, so it is multiple of 3 and 9.
180 is a multiple of 6 because it is divisible by 2 and 3.
The last two digits is 80 and 80 is divisible by 4, so 180 is divisible by 4.
Finally, test for divisibility by 8. The last three digits of 180 are 180, so divide 8 from 180.
180 ÷ 8 = 22 r 4
Answer: {2, 3, 4, 5, 6, 9, 10}

Example 4: If the three-digit number 3A5 is divisible by 3, what is the sum of the possible values of A?

Solution: To test for divisibility by 3, add the digits.
$3 + A + 5 = 8 + A$ = multiple of 3, so 8 + A could be 9, 12, 15 but not 18 because that would make A = 10 and A has to be a single digit.

Set $\{8 + A = 9\}$, $\{8 + A = 12\}$, and $\{8 + A = 15\}$. The resulting values of A are 1, 4, and 7. The sum is 1 + 4 + 7 = **12.**

Answer: 12

Example 5: Is 142,637 divisible by 11?

Solution: According to the rule, add up the even-numbered digits and the odd-numbered digits. The even-numbered digits are 1, 2, 3 and the odd-numbered digits are 4, 6, 7.
1 + 2 + 3 = 6
4 + 6 + 7 = 17
17 - 6 = 11
The positive difference of these two numbers is 11 which means that **142,637 is divisible by 11**.

Answer: 142,637 is divisible by 11

Review Problems

1. Which of these numbers are divisible by 5? {365, 549, 766, 450, 324}

365, 450

2. Is 108 divisible by 6?

Yes

3. How many whole numbers between 1 and 50 are divisible by 3?

50 ÷ 3 = 16 R 2

16 numbers

4. Is 8,715 divisible by 15? Hint: 15 = 5 × 3

Yes

5. What is the largest two-digit whole number that is divisible by 6?

99 ÷ 6 = 16 R 3

16 × 6 = 96

6. What are the remainders when 123,456 is divided by 4, 5, 8, 9, or 10?

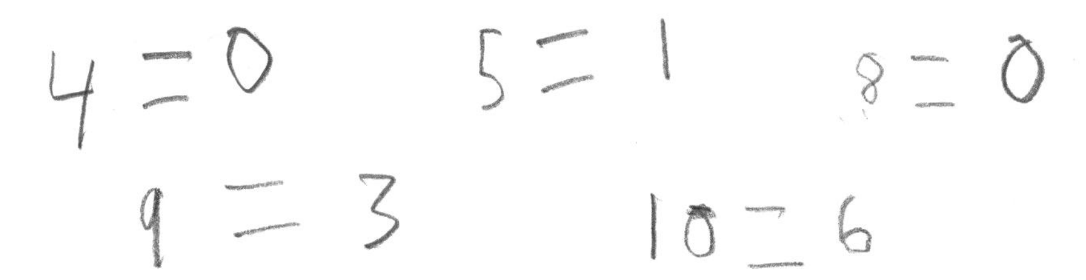

7. How many numbers between 1 and 100 are divisible by 7?

8. If the three-digit number 4A6 is divisible by 9, what is the value of A?

9. Which of these numbers is divisible by 8? {25,412, 34,612, 78,512, 56,388}

78,512

*10. The six-digit number 8A4,B57 is divisible by 11. What is the sum of A and B?

10

11. Jerry is thinking of a number between 1 and 100. The number is a multiple of 11 and is also a multiple of 5. What number is he thinking of?

*12. If 5,A3B is a multiple of 9, what is the sum of all possible values of A + B?

13. What is the remainder when 62,579,304 is divided by 9?

14. How many multiples of 3 are between 11 and 302?

15. The five-digit number A5,2A1 is divisible by 3. If the five-digit number is to be as large as possible, what is the value of A?

16. What is the smallest three-digit number that is a multiple of 5?

17. What is largest two-digit number that is divisible by 7 and 6?

18. Find the remainder when 7,649,432,789 is divided by 9.

19. Marina is trying to remember her favorite number. She remembers that the number is more than 40 and less than 70 and is divisible by 8 and 7. What is her favorite number?

*20. A number is randomly chosen from 1 to 100, inclusive. What is the probability that the number is divisible by 5 and 3? Express your answer as a percent.

Prime Numbers

A **factor** is a positive integer that can be divided evenly into a number.
For example: 5 = 1 × 5, 1 and 5 are factors of 5.

A **prime** number is a positive integer that has 2 distinct factors, 1 and itself.
For example, 2 is a prime number because it only has two factors, 1 and itself (2).

A **composite** number is a positive integer that has 3 or more factors.
For example, 6 is a composite number because it has four factors: {1, 2, 3, 6}.

1 is not a prime number because it only has one factor, 1 or itself.

Fun Fact: There is an infinite number of prime numbers.

Example 1: How many prime numbers are there between 1 and 100?

Solution: One of the ways to find the prime numbers between 1 and 100 is to use the Sieve of Eratosthenes. This is an algorithm that is used to determine and identify prime numbers. We first have to list out the numbers from 1 to 100 in 10 rows and 10 columns.

Step 1: We cross out 1 because it is neither a prime nor a composite number.

~~1~~	2	3	4	5	6	7	8	9	10
11	12	13	14	15	16	17	18	19	20
21	22	23	24	25	26	27	28	29	30
31	32	33	34	35	36	37	38	39	40
41	42	43	44	45	46	47	48	49	50
51	52	53	54	55	56	57	58	59	60
61	62	63	64	65	66	67	68	69	70
71	72	73	74	75	76	77	78	79	80
81	82	83	84	85	86	87	88	89	90
91	92	93	94	95	96	97	98	99	100

Step 2: We cross out the multiples of 2 except 2 because it is a prime number.

~~1~~	2	3	~~4~~	5	~~6~~	7	~~8~~	9	~~10~~
11	~~12~~	13	~~14~~	15	~~16~~	17	~~18~~	19	~~20~~
21	~~22~~	23	~~24~~	25	~~26~~	27	~~28~~	29	~~30~~
31	~~32~~	33	~~34~~	35	~~36~~	37	~~38~~	39	~~40~~
41	~~42~~	43	~~44~~	45	~~46~~	47	~~48~~	49	~~50~~
51	~~52~~	53	~~54~~	55	~~56~~	57	~~58~~	59	~~60~~
61	~~62~~	63	~~64~~	65	~~66~~	67	~~68~~	69	~~70~~
71	~~72~~	73	~~74~~	75	~~76~~	77	~~78~~	79	~~80~~
81	~~82~~	83	~~84~~	85	~~86~~	87	~~88~~	89	~~90~~
91	~~92~~	93	~~94~~	95	~~96~~	97	~~98~~	99	~~100~~

Step 3: We cross out the multiples of 3 except 3 because it is a prime number..

~~1~~	2	3	~~4~~	5	~~6~~	7	~~8~~	~~9~~	~~10~~
11	~~12~~	13	~~14~~	~~15~~	~~16~~	17	~~18~~	19	~~20~~
21	~~22~~	23	~~24~~	25	~~26~~	~~27~~	~~28~~	29	~~30~~
31	~~32~~	~~33~~	~~34~~	35	~~36~~	37	~~38~~	~~39~~	~~40~~
41	~~42~~	43	~~44~~	~~45~~	~~46~~	47	~~48~~	49	~~50~~
~~51~~	~~52~~	53	~~54~~	55	~~56~~	~~57~~	~~58~~	59	~~60~~
61	~~62~~	~~63~~	~~64~~	65	~~66~~	67	~~68~~	~~69~~	~~70~~
71	~~72~~	73	~~74~~	~~75~~	~~76~~	77	~~78~~	79	~~80~~
~~81~~	~~82~~	83	~~84~~	85	~~86~~	~~87~~	~~88~~	89	~~90~~
91	~~92~~	~~93~~	~~94~~	95	~~96~~	97	~~98~~	~~99~~	~~100~~

Step 4: We cross out the multiples of 5 except 5 because it is a prime number..

~~1~~	2	3	~~4~~	5	~~6~~	7	~~8~~	~~9~~	~~10~~
11	~~12~~	13	~~14~~	~~15~~	~~16~~	17	~~18~~	19	~~20~~
21	~~22~~	23	~~24~~	~~25~~	~~26~~	~~27~~	~~28~~	29	~~30~~
31	~~32~~	~~33~~	~~34~~	~~35~~	~~36~~	37	~~38~~	~~39~~	~~40~~
41	~~42~~	43	~~44~~	~~45~~	~~46~~	47	~~48~~	49	~~50~~
~~51~~	~~52~~	53	~~54~~	~~55~~	~~56~~	~~57~~	~~58~~	59	~~60~~
61	~~62~~	~~63~~	~~64~~	~~65~~	~~66~~	67	~~68~~	~~69~~	~~70~~
71	~~72~~	73	~~74~~	~~75~~	~~76~~	77	~~78~~	79	~~80~~
~~81~~	~~82~~	83	~~84~~	~~85~~	~~86~~	~~87~~	~~88~~	89	~~90~~
91	~~92~~	~~93~~	~~94~~	~~95~~	~~96~~	97	~~98~~	~~99~~	~~100~~

Step 5: We cross out the multiples of 7 except 7 because it is a prime number..

~~1~~	2	3	~~4~~	5	~~6~~	7	~~8~~	~~9~~	~~10~~
11	~~12~~	13	~~14~~	~~15~~	~~16~~	17	~~18~~	19	~~20~~
~~21~~	~~22~~	23	~~24~~	~~25~~	~~26~~	~~27~~	~~28~~	29	~~30~~
31	~~32~~	~~33~~	~~34~~	~~35~~	~~36~~	37	~~38~~	~~39~~	~~40~~
41	~~42~~	43	~~44~~	~~45~~	~~46~~	47	~~48~~	~~49~~	~~50~~
~~51~~	~~52~~	53	~~54~~	~~55~~	~~56~~	~~57~~	~~58~~	59	~~60~~
61	~~62~~	~~63~~	~~64~~	~~65~~	~~66~~	67	~~68~~	~~69~~	~~70~~
71	~~72~~	73	~~74~~	~~75~~	~~76~~	~~77~~	~~78~~	79	~~80~~
~~81~~	~~82~~	83	~~84~~	~~85~~	~~86~~	~~87~~	~~88~~	89	~~90~~
~~91~~	~~92~~	~~93~~	~~94~~	~~95~~	~~96~~	97	~~98~~	~~99~~	~~100~~

We only need to check the prime numbers that are less than or equal to the square root of 100. $\sqrt{100} = 10$. Since 2, 3, 5 and 7 are the only prime numbers less than or equal to 10, we are done and the remaining squares are prime numbers.

Answer: **There are 25 prime numbers between 1 and 100.**

Tip: It is more important to understand the process of using the Sieve rather than memorizing it.

Example 2: What are the prime factors of 12?

Solution:
12 = 3 × 2 × 2

Answer: 2 and 3 are the prime factors of 12.

Example 3: Is 101 a prime number? If not, find its prime factors.

Solution: Using the Sieve of Eratosthenes, the only prime numbers that need to be checked are up to the square-root of 101 ($\sqrt{101} < 11$). This means that only 2, 3, 5 and 7 need to be checked because 11 × 11 = 121 which is greater than 101.

101 ÷ 2 = 50 R 1
101 ÷ 3 = 33 R 2
101 ÷ 5 = 20 R 1
101 ÷ 7 = 14 R 3

Thus, **101 is a prime number** because it is not divisible by 2, 3, 5 or 7.

Answer: Yes, 101 is a prime number

Example 4: What prime factors do 15 and 20 share?

Solution:
15 = 5 × 3
20 = 5 × 2 × 2

15 and 20 share 5 as a prime factor.

Answer: 5

Example 5: Find the sum of all prime numbers less than 100 that end in 3.

Solution: The numbers that end in 3 are 3, 13, 23, 33, 43, 53, 63, 73, 83, and 93.

2: None of the numbers are divisible by 2.
3: 3, 33, 63, and 93 are divisible by 3 but 3 is a prime number.
5: None of the numbers are divisible by 5.
7: 63 is divisible by 7.

Thus, 3 + 13 + 23 + 43 + 53 + 73 + 83 = **291**.

Answer: 291

Review Problems

1. What is the sum of the first 4 prime numbers?

2. Is 84 a prime number? If not, find its prime factors.

3. What is the largest prime number less than 100?

4. How many prime numbers are divisible by 2?

5. What are the prime factors of 40?

6. Is 119 a prime number? If not, find its prime factors. Hint: $121 = 11 \times 11$.

7. Is 163 a prime number? If not, find its prime factors. Hint: $169 = 13 \times 13$.

8. What prime numbers end in 5?

9. How many prime numbers are between 1 and 50?

10. Find the sum of the prime numbers between 20 and 50.

11. What prime factors do 60 and 36 share?

12. Find the sum of all prime numbers less than 100 that end in 7.

*13. What is the largest prime number less than 190 that ends in 7?

14. What prime factors do 135 and 90 share?

15. Find the sum of all prime numbers less than 100 that end in 1.

*16. Is 1001 a prime number? If not, find its prime factors.

*17. Find the prime factors of 183,183. Hint: 183,183 = 183 × 1001

18. What is the product of the first three prime numbers?

*19. What is the greatest prime number less than 200? Hint: Work your way down from the greatest **odd** number and check if it's a prime number.

*20. Which of the numbers between 100 and 110 are prime numbers?

Prime Factorization I

Prime Factorization

The **prime factorization** of a number is the unique set of prime numbers that, when multiplied, make up the number. For example, the prime factorization of $30 = 2 * 3 * 5$ where 2, 3, and 5 are prime numbers.

If there are repeats of a prime factor in a number, it can be denoted with a superscript number showing how many times to multiply that number. For example, the prime factorization of $72 = 2^3 * 3^2$ where 2 and 3 are prime numbers. The superscript is called the **exponent**. The exponent in 2^3 is 3, which is the number of times that 2 is multiplied. Thus, it can be rewritten as $72 = 2^3 * 3^2 = 2 * 2 * 2 * 3 * 3$.

Note: Any positive integer with an exponent of 0 is equal to 1. (For example: $5^0 = 1$)

Number of Factors

To find the number of factors in a number, first, find the prime factorization for that number.

Let's look at a number with one prime factor:
$N = p^n$ where p is a prime number and n is the exponent.
The number of factors is $(n + 1)$.

Now, let's examine a number with two prime factors:
$N = p_1^{n_1} p_2^{n_2}$ where p_1, p_2 are prime numbers and n_1, n_2 are their exponents.
The number of factors is $(n_1 + 1) * (n_2 + 1)$.

General Rule: To find the number of factors of a number, add one to each exponent and calculate the product.

Sum of Factors

To find the sum of factors in a number, first, find the prime factorization for that number.

Let's look at a number with one prime factor:
$N = p^n$ where p is a prime number and n is the exponent.
The sum of factors is $(p^0 + p^1 + ... + p^n)$.

Now, let's examine a number with two prime factors:
$N = p_1^{n_1} p_2^{n_2}$ where p_1, p_2 are prime numbers and n_1, n_2 are their exponents.
The sum of factors is $(p_1^0 + p_1^1 + ... + p_1^{n_1}) * (p_2^0 + p_2^1 + ... + p_2^{n_2})$.

General Rule: To find the sum of factors in a number, find the sum of the powers of each prime number, and then calculate the product of these sums.

Example 1: Find the prime factorization for 13.

Solution: 13 is a prime number, therefore 13 is the prime factorization.

Answer: $13 = 13$

Example 2: Find the prime factorization for 36.

Solution: 36 = 6 × 6 = 3 × 2 × 3 × 2 = 3 × 3 × 2 × 2

Answer: $36 = 2^2 * 3^2$

Example 3: Find the number of factors for 36.

Solution: Using $36 = 2^2 * 3^2$,
the rule is to add one to each exponent and calculate the product.

$(2 + 1) * (2 + 1) = 3 * 3 = 9$

Answer: **9 factors**

Example 4: Find the sum of all the factors for 36.

Solution: Using $36 = 2^2 * 3^2$,
the rule is to add up all the powers of 2 and all the powers of 3 that are prime factors of 36 and calculate the product of the two sums.
$(2^0 + 2^1 + 2^2) * (3^0 + 3^1 + 3^2) = (1 + 2 + 4) * (1 + 3 + 9) = 7 * 13 = 91$

Answer: 91

Example 5: Find the number of factors and sum of all the factors for 90.

Solution:
Find the prime factorization for 90: 90 = $2^1 3^2 5^1$

Number of factors: $(1 + 1) * (2 + 1) * (1 + 1) = 2 * 3 * 2 = 12$

Sum of factors: $(2^0 + 2^1) * (3^0 + 3^1 + 3^2) * (5^0 + 5^1) = 3 * 13 * 6 = 234$

Answer:
Number of factors: 12
Sum of factors: 234

Review Problems

1. Find the prime factorization for 17.

2. Find the prime factorization for 91.

3. Find the number of factors for 6.

4. Find the number of the factors for 65.

5. Find the sum of all the factors for 30.

6. Find the sum of all the factors for 77.

7. Find the number of the factors for 84.

8. Find the prime factorization for 60.

9. How many more factors does 60 have than 30?

10. Find the prime factorization for 8000.

11. Find the prime factorization and the number of factors for 480.

12. Find the sum of all the factors for 8, 16, 32 and 64. List each sum separately.

13. Find the number of factors for 630.

14. Find the sum of the factors that are not factors of 30 but are factors of 60.

15. Find the number of factors for $13^9 31^4$.

16. How many factors do $2^{14}5^{7}11^{5}$ and $3^{8}7^{3}13^{12}$ share?

17. Find the number of factors for $11^{3}13^{1}17^{14}$.

18. Find the positive difference between the number of factors and the sum of all the factors for 210.

19. How many factors does 42 and 630 share?

20. Find the number of the factors for 12,000.

Prime Factorization II

Perfect Square

If all the exponents in a prime factorization are multiples of 2, the number is a perfect square. For example, $100 = 2^2 * 5^2$ implies that 100 is a perfect square.

Perfect Cube

If all the exponents in a prime factorization are multiples of 3, the number is a perfect cube. For example, $216 = 2^3 * 3^3$ implies that 216 is a perfect cube.

Even and Odd Factors

All prime numbers other than 2 are odd numbers. Thus, the only prime factor that makes a number even is 2. To calculate the number of even factors, leave out 2^0 or in other words, do not add 1 to the exponent of 2.
To calculate the number of odd factors, disregard the powers of 2. Find the number of factors for the number excluding the factors of 2.

Example 1: Find the number of factors that are perfect squares for 36.

Solution: Using $36 = 2^2 * 3^2$, the exponents of 2 that are perfect squares are $2^0, 2^2$ (2) and the exponents of 3 that are perfect squares are $3^0, 3^2$ (2). Thus, we multiply the number of each exponent with perfect square to find the number of factors that are perfect squares.
$2 * 2 = 4$

Answer: **4 factors**

Example 2: Find the number of even factors for $2^2 3^4 5^3$.

Solution: Examining the powers of 2, only 2^0 = 1 is an odd factor. Therefore, when counting the number of factors, leave out adding 1 to the exponent of 2. Thus,
$(2) * (4 + 1) * (3 + 1) = 40$

Answer: 40 even factors

Example 3: Find the number of odd factors for $2^2 3^4 5^3$.

Solution: Disregarding the exponents of 2, the problem can be simplified to finding the factors of $3^4 5^3$, which are all odd factors. $(4+1)*(3+1)=20$

Answer: 20 odd factors

Example 4: Find the sum of all the even factors for $2^2 3^3 5^1$.

Solution: Disregarding 2^0 in the sum of the powers of 2:
$(2^1+2^2)*(3^0+3^1+3^2+3^3)*(5^0+5^1)=$
$(2+4)*(1+3+9+27)*(1+5)=6*40*6=1440$

Answer: 1440

Example 5: How many factors are divisible by 3 but not divisible by 9 for $2^4 3^4 5^4$.

Solution: The only power of 3 that is divisible by 3 but not divisible by 9 is 3^1.

Thus, there is only one possibility for the exponent of 3s:
$(4+1)*(1)*(4+1)=5*1*5=25$

Answer: 25 factors

Example 6: Find the sum of all the factors that are divisible by 15 for $2^2 3^3 5^1$.

Solution: 15 = 3 × 5 or $3^1 5^1$.

Thus, the exponent for the powers of 3 and 5 must be at least one.
$(2^0+2^1+2^2)*(3^1+3^2+3^3)*(5^1)=$
$(1+2+4)*(3+9+27)*(5)=7*39*5=1365$

Answer: 1365

Review Problems

1. Find the number of even factors for $2^5 3^8 7^4$.

2. Find the number of odd factors for $2^5 3^8 7^4$.

3. How many perfect squares are factors of $2^5 3^8 7^4$?

4. How many factors are divisible by 3 for $3^{10} 7^5 13^2$?

5. How many perfect cubes are factors of $2^4 3^9 11^7$?

6. How many perfect squares are between 1 and 100, inclusive?

7. How many perfect cubes are between 1 and 100, inclusive?

8. Is 64 a perfect square? A perfect cube?

9. Find the number of factors that are divisible by 6 for $2^2 3^4 5^3$.

10. Find the sum of factors that are divisible by 2 but not divisible by 3 for $2^2 3^4 5^3$.

11. Find the number of factors that are perfect squares for $2^2 3^4 5^3$.

12. Find the sum of all the factors that are perfect squares for $2^2 3^3 5^3$.

13. Find the number of factors that are perfect cubes for $2^2 3^4 5^3$.

14. Find the number of even factors for $2^2 3^4 5^2 7^4$.

15. Find the number of odd factors for $2^2 3^4 5^2 7^4$.

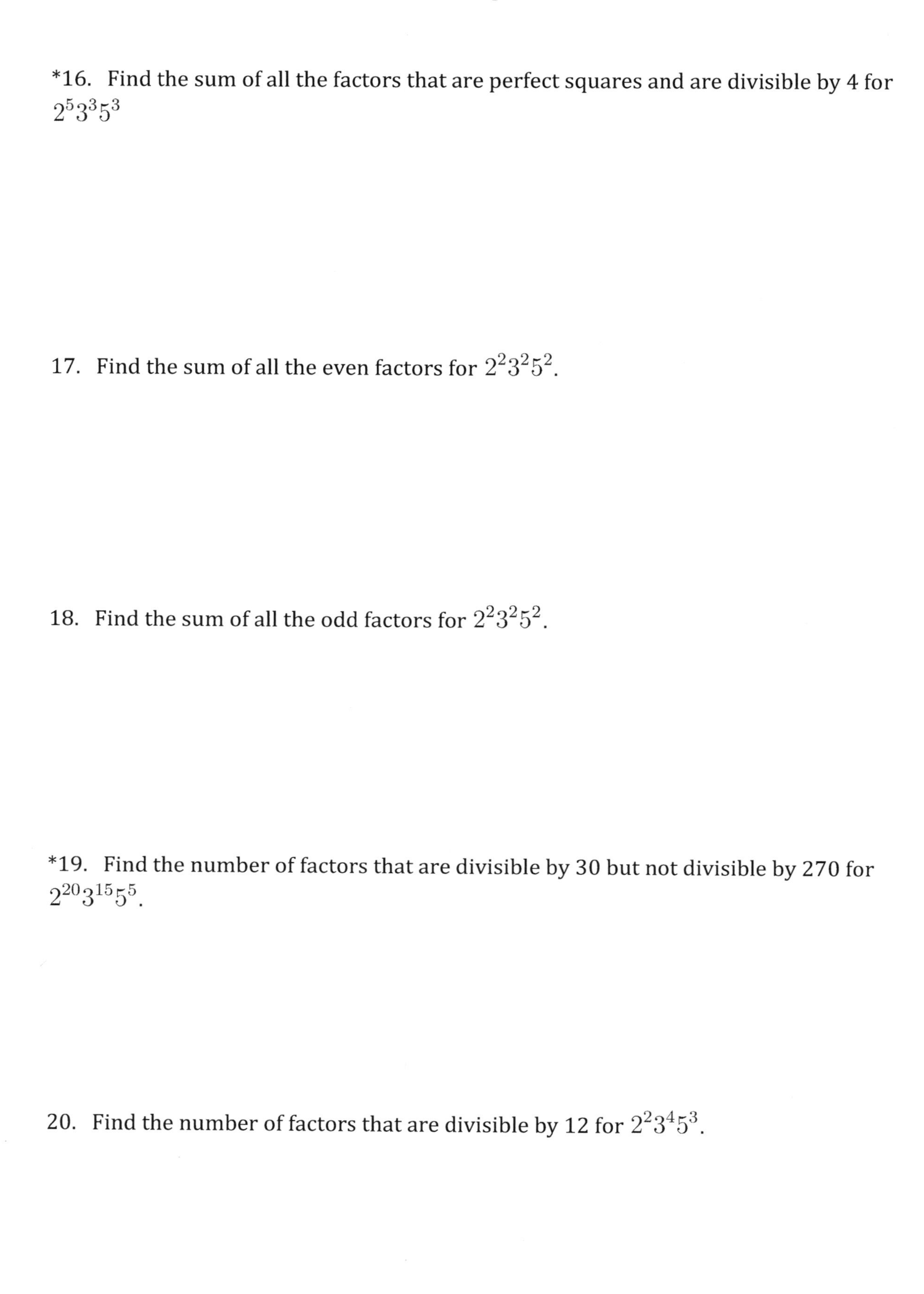

*16. Find the sum of all the factors that are perfect squares and are divisible by 4 for $2^5 3^3 5^3$

17. Find the sum of all the even factors for $2^2 3^2 5^2$.

18. Find the sum of all the odd factors for $2^2 3^2 5^2$.

*19. Find the number of factors that are divisible by 30 but not divisible by 270 for $2^{20} 3^{15} 5^5$.

20. Find the number of factors that are divisible by 12 for $2^2 3^4 5^3$.

Money

In the United States, we use coins and paper monies to purchase goods. The name of the currency is called the US dollar denoted by $. In a dollar, there are 100 cents which is denoted by ¢.

Different Coins and their values:
Penny = 1¢
Nickel = 5¢
Dime = 10¢
Quarter = 25¢
Half-dollar = 50¢

Dollar Conversions (Dollar = 100¢)
Dollar = 100 pennies
Dollar = 20 nickels
Dollar = 10 dimes
Dollar = 4 quarters
Dollar = 2 half-dollars

For $8.43, it can be read as "$8 and 43¢" or "8 dollars and 43 cents".

Example 1: Janet bought a bag of apples at a grocery store for $10.41. She received $4.59 in change. How much did she give the cashier? Express your answer in dollars.

Solution:
Price of Item + Change = Amount Given
$10.41 (price of the apples) + $4.59 (change) = $15.00 (amount given)

Answer: $15 or $15.00

Example 2: Carl gave the cashier a $50-bill to pay for a toy car and received $11.18 in change. How much did the toy car cost? Express your answer in dollars.

Solution:
Price of Item = Amount Given – Change
Price of Toy Car = $50.00 (amount given) - $11.18 (change) = $38.82

Answer: $38.82

Example 3: A store sells movies in DVD's for \$18.00 each or a package of 4 for \$65.00. How much would one save by buying 12 DVDs in packages rather than buying 12 DVDs individually?

Solution:
12 DVDs individually: $12 \times \$18.00 = \$216.00$

12 DVDs in packages of four: $12 \div 4 = 3$ packages of four
$3 \times \$65.00 = \$195.00$
$\$216.00 - \$195.00 = \$21.00$

Answer: \$21 or \$21.00

Example 4: A store sells erasers with three options: 1 for 5 cents, a pack of 4 for 18 cents or a pack of 12 for 50 cents. James needs 82 erasers. In order to save money, what is the least amount that he needs to pay?

Solution:
First, compare the prices for the three options.

1 for 5 cents can be multiplied by 12:	12 for 60 cents
A pack of 4 for 18 cents can be multiplied by 3:	12 for 54 cents
A pack of 12 for 50 cents:	12 for 50 cents

Clearly, a pack of 12 for 50 cents is the cheapest option, a pack of 4 for 18 cents is the second cheapest and the 1 for 5 cents is the most expensive.

$82 \div 12 = 6$ R 10
James can buy 6 packs of 12 but needs 10 more. He should buy as many packs of 4 because it is the second cheapest option.

$10 \div 4 = 2$ R 2
James can buy 2 packs of 4 but needs 2 more. Finally, he can buy 2 individually to get the total of 82 erasers that he needs.

In all, he needs 6 packs of 12, 2 packs of 4 and 2 individual erasers.
Now, it can be totalled:
$6 \times 50 + 2 \times 18 + 2 \times 5 = 300 + 36 + 10 = 346¢$ or \$3.46

Answer: \$3.46

Example 5: How many ways are there to make 10 cents using pennies, nickels and/or dimes?

Solution:

To solve this, draw a table and start with the maximum number of dimes.

The maximum is 1 dime which has 1 case (0 pennies, 0 nickels, 1 dime).
The next case to consider is 0 dimes and the maximum number of nickels which is 2 nickels.
Working down, there are 3 cases (2 nickels, 1 nickel, 0 nickels).

Thus, there are 4 total cases.

Pennies (1¢)	Nickels (5¢)	Dimes (10¢)	Total (10¢)
0	0	1	10¢
0	2	0	10¢
5	1	0	10¢
10	0	0	10¢

Answer: 4 ways

Chapter 6

Review Problems

1. Amy bought a vanilla ice cream cone at the county fair for $6.87 and received $2.56 in change. How much money did she give the cashier? Express your answer in dollars.

2. Josh gave the cashier $1,124.89 to pay for a laptop computer which cost $1,089.39. How much money should he receive in change? Express your answer in dollars.

3. Hannah bought 3 pencils, which cost 28 cents each, with a one-dollar bill. How much did she receive in change? Express your answer in cents.

4. Sasha has 3 half-dollars, 8 quarters, 11 dimes, 7 nickels and 13 pennies in her pink piggy bank. How much money does she have in total?

5. Ross has quarters in 4 stacks with each stack having 8 quarters and dimes in 5 stacks with each stack having 7 dimes. How much money does he have in total? Express your answer in dollars.

6. How many ways are there to make 15 cents using pennies and/or nickels?

7. The tickets to a concert are priced at $154 for two or $83 for one. How much would Cassie and Jacob save if they bought the tickets for two instead of paying individually for each ticket? Express your answer in dollars.

8. At a convenience store, pens sell for 99 cents each, pencils sell for 49 cents each and erasers sell for 39 cents each. If Riley needs 2 pens, 3 pencils and 5 erasers, how much money does he need? Express your answer in dollars.

9. At the candy store, Alex bought 4 chocolate bars costing $2 each and 5 lollipops costing 50 cents each. If Alex had a $20-bill in his wallet, how much money did he have left after purchasing the chocolate bars and lollipops? Express your answer in dollars.

10. 3 adults and 2 senior citizens take 4 children to see a movie. If one adult ticket cost $12, one senior citizen ticket cost $10 and one child ticket cost $7.50, how much was the total cost for all the tickets?

11. Rank these deals from least expensive to most expensive for one sunglasses: 1 sunglasses for $14, 3 sunglasses for $40 or 9 sunglasses for $124.

12. A store sells boxes with two options: 1 for $3.00 or 11 for $30.00. What is the least amount that John needs to spend to get exactly 100 boxes?

13. How many ways are there to make 17 cents using pennies and/or nickels?

14. Julie takes a taxi from her house to the airport. The taxi charges $4.50 to start the meter and $1.10 for each mile traveled. If the distance is 15 miles from Julie's house to the airport, how much will she have to pay the taxi driver when she arrives at the airport? Express your answer in dollars.

15. A company sells shirts with three different options: 2 for $19, 5 for $45 or 20 for $170. If Logan needs to purchase exactly 132 shirts, what is the least amount he has to pay?

16. How many ways are there to make 25 cents using quarters, dimes and/or nickels?

17. A phone company charges $15 a month for their services and charges an extra 50 cents per minute for long distance calls. If Josie's July bill is $99, how many minutes did she call long distance?

18. 4 adults and 2 children went to visit an aquarium. If one adult ticket cost $14 and one child ticket cost $5, how much was the total cost for all the tickets?

19. Ten adults and ten children go to see a movie at the theatre. If the total amount came out to be $210 and an adult ticket cost $14, how much did a child ticket cost?

*20. How many ways are there to make $20 using $10-bills, $5-bills and/or $1-bills?

Exponents and Roots

In the prime factorization chapters, we explored exponents, perfect squares and perfect cubes. This chapter will be an extension of those topics.

Let's look at 3^4. 3 is the **base number** and 4 is the **exponent**. The exponent denotes the number of times the base number is multiplied. Thus, $3^4 = 3 \times 3 \times 3 \times 3 = 81$.

When two numbers are multiplied and they have the same base number, keep the base number and add the exponents. Use this formula: $a^b * a^c = a^{b+c}$
For example: $3^4 \times 3^5 = 3^{(4+5)} = 3^9$

If the exponent is 2, the base number is said to be "squared" or "raised to the second power". For example: 2^2 is read as "2 squared" or "2 to the second power".

If the exponent is 3, the base number is said to be "cubed" or "raised to the third power".

Exponents can be integers but they can also be fractions. This introduces roots ($\sqrt{\ }$).

Let's look at a square root: $\sqrt{4}$ or $\sqrt[2]{4}$ (Note: $\sqrt{\ } = \sqrt[2]{\ }$). Use this formula: $\sqrt[c]{a^b} = a^{\frac{b}{c}}$
The base number is 4 but the exponent will be a fraction of ½: $\sqrt[2]{4^1} = 4^{\frac{1}{2}}$

The 2 on the outside of the root sign denotes how many times a number must be multiplied to yield the base number.

Since $4 = 2^2 = 2 \times 2$, $\sqrt[2]{4} = 2$.

It can also be written as $\sqrt[2]{2^2} = 2^{\frac{2}{2}} = 2^1 = 2$

Formulas

$$a^0 = 1 \text{ where } a \neq 0 \qquad a^b * a^c = a^{b+c} \qquad a^b \div a^c = a^{b-c}$$

$$(a^b)^c = a^{b*c} = a^{bc}$$

$$a^1 = a \qquad a^{-b} = \frac{1}{a^b} \qquad a^b = \frac{1}{a^{-b}}$$

$$\sqrt[c]{a^b} = a^{\frac{b}{c}} \qquad \sqrt[c]{a_1^{b_1} a_2^{b_2}} = \sqrt[c]{a_1^{b_1}} * \sqrt[c]{a_2^{b_2}}$$

First 10 Square Roots

$$\sqrt{1} = 1 \qquad \sqrt{6} \approx 2.449$$
$$\sqrt{2} \approx 1.414 \qquad \sqrt{7} \approx 2.646$$
$$\sqrt{3} \approx 1.732 \qquad \sqrt{8} \approx 2.828$$
$$\sqrt{4} = 2 \qquad \sqrt{9} = 3$$
$$\sqrt{5} \approx 2.236 \qquad \sqrt{10} \approx 3.162$$

Example 1: Simplify $2^5 \times 4^7 \times 8^3$ as a power of 2.

Solution: The base numbers are $2 = 2^1$, $4 = 2^2$ and $8 = 2^3$.

$2^5 \times 4^7 \times 8^3 = 2^5 \times (2^2)^7 \times (2^3)^3 = 2^5 \times 2^{14} \times 2^9 = 2^{(5+14+9)} = 2^{28}$

Answer: $\mathbf{2^{28}}$

Example 2: Simplify $\sqrt{72}$.

Solution: $72 = 2^3 3^2$

Therefore, $\sqrt{72} = \sqrt{2^3 3^2}$

The largest factor of 72 that is a perfect square is $2^2 3^2$ where the exponents are factors of 2.

$\sqrt{72} = \sqrt{2^3 3^2} = \sqrt{2^2 3^2} * \sqrt{2^1} = 2^{\frac{2}{2}} 3^{\frac{2}{2}} \sqrt{2^1} = 2 * 3 * \sqrt{2} = 6\sqrt{2}$

Answer: $6\sqrt{2}$

Example 3: Find the two consecutive integers that are closest to $\sqrt{54}$.

Solution:

The square root of 54 must be between 7 and 8. $\sqrt{49} < \sqrt{54} < \sqrt{64}$

$(7 < \sqrt{54} < 8)$

Answer: 7 and 8

Example 4: Simplify $\sqrt{\sqrt[3]{64}}$. Hint: Simplify the most nested root first.

Solution: 64 = 2^6

$\sqrt[3]{2^6} = 2^{\frac{6}{3}} = 2^2$ $\qquad$ $\sqrt{\sqrt[3]{64}} = \sqrt{2^2} = 2^{\frac{2}{2}} = 2^1 = 2$

Or

$$((2^6)^{\frac{1}{3}})^{\frac{1}{2}} = ((2^{\frac{6}{3}})^{\frac{1}{2}} = (2^2)^{\frac{1}{2}} = 2^{\frac{2}{2}} = 2^1 = 2$$

Answer: 2

Example 5: Find the largest perfect square less than 1,000.

Solution: $\sqrt{1000} = \sqrt{10^3} = \sqrt{10^2} * \sqrt{10^1} = 10\sqrt{10}$

From the approximation of $\sqrt{10} = 3.162$,
$10\sqrt{10} = 10 * 3.162 = 31.62$

Thus, 31^2 = 961 is the largest perfect square less than 1,000.

Answer: 961

Review Problems

1. Simplify $5^{13} \times 25^{7}$ as a power of 5.

2. Simplify $7^{6} \div 7^{(-4)} \div 7^{10}$.

3. Simplify $\dfrac{1}{8^{-2}}$.

4. Simplify $\sqrt{100}$.

5. Simplify $\sqrt[3]{72}$.

6. Simplify $\sqrt{2^{32}} * \sqrt{2^{-32}}$.

7. Simplify $(\sqrt[7]{384})^0$.

8. Find the integer portion of the decimal representation of $\sqrt{19}$.

9. Find the two consecutive integers that are closest to $\sqrt[3]{150}$.

10. Simplify $\sqrt[14]{81^7}$.

11. Simplify $\sqrt[6]{2^{14}} * \sqrt[3]{2^5}$.

12. Simplify $\sqrt[3]{2^2 11^1} * \sqrt[6]{2^2 11^4}$.

13. Simplify $\sqrt[11]{\sqrt[3]{(8^4)^7 * 32^3}}$.

14. Simplify $(\sqrt[3]{27000000})^2$

15. Simplify $\sqrt[3]{\sqrt[3]{\sqrt[2]{9^9}}}$.

16. Find the largest perfect square less than 500.

17. How many perfect squares are between 100 and 1,000, inclusive?

18. Find the largest perfect square less than 9,000.

19. Simplify $\sqrt[2]{11} * \sqrt[3]{11} * \sqrt[6]{11}$.

20. Simplify $\sqrt[9]{5^2} * \sqrt[18]{5^2} * \sqrt[3]{5^8}$.

Fractions and Percentages

A **fraction** is the ratio of two numbers. It can be written with two numbers separated by a line. The number on the top is called the **numerator** and the number on the bottom is called the **denominator**.

$$\text{Fraction} = \frac{\text{Numerator}}{\text{Denominator}} = \text{Numerator} \div \text{Denominator}$$

Types of Fractions

Proper Fraction: A fraction less than 1. In other words, the numerator is less than the denominator.

Examples: $\frac{1}{2}$, $\frac{2}{3}$, $\frac{6}{11}$

Improper Fraction: A fraction greater than 1. In other words, the numerator is greater than the denominator.

Examples: $\frac{4}{3}$, $\frac{9}{7}$, $\frac{14}{11}$

Mixed Numbers: A number with a non-zero integer in front and a proper fraction at the end.

Examples: $3\frac{1}{2}$, $5\frac{5}{6}$, $4\frac{1}{13}$

Converting a Mixed Number to an Improper Fraction

$$2\frac{4}{7} = \frac{(2 * 7) + 4}{7} = \frac{18}{7}$$

Step 1) Multiply the denominator by the integer in front
Step 2) Add the product to the numerator
Step 3) Place the sum over the denominator

Converting an Improper Fraction to a Mixed Number

$\frac{37}{5}$ can be rewritten as 37 ÷ 5 = 7 R 2

$$\frac{37}{5} = 7\frac{2}{5}$$

Step 1) Find the quotient and the remainder when the numerator is divided by the denominator
Step 2) Place the quotient as the integer in front of the fraction
Step 3) Place the remainder over the denominator as the proper fraction

Percentages

A **percentage** is expressed as a number over 100.

For example, $\frac{75}{100}$ is equivalent to 75%

Converting Fractions to Percentages

$\frac{2}{5} = 0.4$ $\quad 0.4 * 100 = 40$ $\quad \frac{2}{5} = 40 \text{ percent}$ = 40%

Convert the fraction into a decimal and then multiply by 100.

Converting Percentages to Fractions

80% = $80 \text{ percent} = \frac{80}{100} = \frac{4}{5}$

Place the number in front of the percent sign over 100, and reduce the fraction if necessary.

Conversion Table (Fraction to Decimal to Percent):

Fraction	Decimal	Percent
0	0	0%
1	1	100%
½	0.50	50%
⅓	$0.\overline{3}$	$33.\overline{3}\%$
⅔	$0.\overline{6}$	$66.\overline{6}\%$
¼	0.25	25%
¾	0.75	75%
⅕	0.2	20%
⅖	0.4	40%
⅗	0.6	60%
⅘	0.8	80%

Note: The line above the 3 in $0.\overline{3}$ denotes that the 3 repeats infinitely after the decimal point. ($0.\overline{3}$ = 0.333333...)

Example 1: Find the sum of the two proper fractions $\frac{1}{2}$ and $\frac{5}{7}$. Express your answer as a mixed number.

Solution: To add two fractions, the denominator must be the same. This can be done by multiplying a fraction equal to 1 (For example: $\frac{7}{7} = 1$)

$$\frac{1}{2} * \frac{7}{7} + \frac{5}{7} * \frac{2}{2} = \frac{7}{14} + \frac{10}{14} = \frac{17}{14}$$

17 ÷ 14 = 1 R 3

$$\frac{17}{14} = 1\frac{3}{14}$$

Answer: $1\frac{3}{14}$

Example 2: Reduce the fraction $\frac{180}{100}$. Express your answer as an improper fraction.

Solution: A reduced fraction is where the Greatest Common Factor of the numerator and the denominator is 1. This can be found by dividing a fraction equal to 1 until the numerator and the denominator share no prime factors. Finding the prime factorization of the numbers is a good place to start.

$$\frac{180}{100} = \frac{2^2 3^2 5^1}{2^2 5^2}$$

The numerator and the denominator share $2^2 5^1 = 20$ as a factor.

$$\frac{180 \div 20}{100 \div 20} = \frac{9}{5}$$

The reduced fraction is $\frac{9}{5}$.

Answer: $\frac{9}{5}$

Example 3: Solve: $\frac{3}{11} \div \frac{45}{44}$

Solution: When a fraction is divided by another fraction, it can be rewritten as the product of the first fraction and the **reciprocal** of the second fraction.

$$\frac{3}{11} \div \frac{45}{44} = \frac{3}{11} * \frac{44}{45} = \frac{2^2 3^1 11^1}{3^2 5^1 11^1}$$

Factors shared on the top and the bottom of the fraction can be canceled out.

Thus, $\frac{2^2}{3^1 5^1} = \frac{4}{15}$

Answer: $\frac{4}{15}$

Example 4: What percent of 68 is 17?

Solution: The original fraction can be reduced to ¼.

$$\frac{17}{68} * 100 = \frac{1}{4} * 100 = 25 \text{ percent}$$

= 25%

Answer: 25%

Example 5: A picnic table at a store is selling for $150. Jason has a 20% off coupon. How much would he have to pay if he uses the coupon? Express your answer in dollars.

Solution: 20% off means that Jason had to pay (100 – 20)% = 80% of the original price.

$$\frac{80}{100} * 150 = \frac{4}{5} * 150 = 4 * 30 = 120$$

Answer: $120

Chapter 8

Review Problems

1. Express $\frac{15}{20}$ as a reduced fraction.

2. Solve: ½ + ⅓ + ¼. Express your answer as a reduced mixed number.

3. Convert $\frac{5}{4}$ to a decimal and a percent.

4. Solve: $\frac{100}{120} * \frac{4}{14} * \frac{35}{8}$. Express your answer as an improper fraction.

5. What percent of 100 is 35?

6. What percent of 25 is 6?

7. Solve: $\frac{13}{12} \div \frac{39}{48}$. Express your answer as a reduced mixed number.

8. What percent of 40 is 28?

9. Herald answered 80% of the 120 questions on his test correctly. How many questions did he get wrong?

10. Arnold, Bob and Carl are sharing a pizza. Arnold ate $\frac{1}{6}$ of the pizza, Bob ate $\frac{1}{4}$ of the pizza and Carl ate the rest. What fraction of the pizza did Carl eat? Express your answer as a reduced proper fraction.

*11. If there are 15 boys and 25 girls in a book club, what percent of the book club members are boys?

12. 800 students attend a high school. If the enrollment increases by 20% the next year, how many students will attend the high school next year?

13. If there are 8 children and 12 adults at the park, what percent of the people at the park are adults?

14. Sandra bought five shirts at a clothing store. If the price of each shirt was $25 and she used a 40% off coupon that could only be used on one shirt, how much did she pay for the five shirts?

15. A smartphone is selling for $800 in January. One month later, in February, the phone was 20% off the original price. In the following month, March, the phone's price increased by 20% from February. How much was the phone's price in March?

16. Barbara has a plastic container that consists of red, blue and green marbles. If there are 200 marbles in the container in which 25% are green, 20% are red and the remaining marbles are blue, how many marbles are blue?

17. There are 64 students consisting of 36 boys and 28 girls in Mr. Smith's math class. If ¼ of the class is absent one day, and ⅓ of the boys are absent, how many girls are absent that day?

18. On an exam with 275 problems, Ruth answered 76% of the questions correct, left 4% of the questions blank and got the remaining questions incorrect. If she receives 1 point for every correct answer, 0 points for every blank answer and loses ⅕ of a point for every incorrect answer, what was her score on the test?

*19. A laptop's price increased by 25%. What percent must the laptop's new price be reduced in order to get back to the original price?

20. Joshua has 96 jellybeans. He gives ⅓ of the jellybeans to his friend, Paul. Then he gives ¼ of the jellybeans he has left to his brother, George. And finally, he gives ⅙ of the remaining jellybeans to his sister, Nancy. How many jellybeans does Joshua have left at the end?

Chapter 9

Proportions and Ratios

Ratio of X to Y = X to Y = $X : Y$ = $\frac{X}{Y}$

If $\frac{A}{B} = \frac{C}{D}$ where $B \neq 0, D \neq 0$, then $AD = BC$.

Odds are used to express the chance of something occurring or not occurring.
Odds for or in favor = Ratio of Number of Successes to Number of Failures
Odds against or not in favor = Ratio of Number of Failures to Number of Successes

Thus, the probability of something occurring is $\frac{\text{Successes}}{\text{Successes} + \text{Failures}}$

Example 1: There are 20 students in a class. The ratio of boys to girls is 2 to 3. Find the number of boys and girls in the class.

Solution: When solving proportion and ratio problems, think of the numbers in the ratio as parts of a whole, in which the whole is the total number of parts. Therefore, there are 5 parts total (2 + 3) which means that 2 out of 5 students would be boys and 3 out of 5 students would be girls.

This means that ⅖ of the total number of students are boys and ⅗ of the total number of students are girls.

$$\text{Boys} = \frac{2}{5} * 20 = 8$$

$$\text{Girls} = \frac{3}{5} * 20 = 12$$

Boys + Girls = 8 + 12 = 20 which is the total number of students.

Answer: 8 boys and 12 girls

Example 2: The odds of Jeff hitting the target with his bow and arrow is 4 to 1. What are the chances that he misses the target? Express your answer as a percent.

Solution: The odds are in favor, thus the ratio is 4 successes to 1 failure or 4 to 1.

The probability that he misses the target is: $\frac{\text{Failures}}{\text{Successes} + \text{Failures}} = \frac{1}{4+1} = \frac{1}{5}$

$\frac{1}{5} * 100 = 20 \text{ percent}$ = 20%

Answer: 20%

Example 3: The ratio of red balls to green balls to blue balls is 8:3:4. There are 300 balls in total. If 100 red balls and 20 blue balls are removed, what is the new ratio of red balls to green balls to blue balls?

Solution: 8 + 3 + 4 = 15. Therefore, there are 15 parts that make a whole.

$$\text{Red} = \frac{8}{15} * 300 = 8 * 20 = 160$$
$$\text{Green} = \frac{3}{15} * 300 = \frac{1}{5} * 300 = 60$$
$$\text{Blue} = \frac{4}{15} * 300 = 4 * 20 = 80$$

If 100 red balls are removed, there are (160 – 100) = 60 red balls remaining.
If 20 blue balls are removed, there are (80 - 20) = 60 blue balls remaining.

There are 60 green balls, therefore, the number of red, green and blue balls are the same. Thus, the ratio is (60 to 60 to 60) or (1 to 1 to 1) or 1:1:1.

Answer: 1 to 1 to 1 or 1:1:1

Example 4: The ratio of the number of cars to trucks in a parking lot is 13 to 12. If there are 6 more cars than trucks, find the number of cars and the number of trucks in the parking lot.

Solution: The difference in the number of parts for cars and trucks is 13 – 12 = 1. This means that for every 13 cars, there is 1 more car than trucks.
Thus, if the factor multiplied to the ratio is 6, the difference in cars and trucks would be 6.

$$\text{Cars} = 13 * 6 = 78$$
$$\text{Trucks} = 12 * 6 = 72$$

Sure enough, 78 – 72 = 6.

Answer: 78 cars and 72 trucks

Example 5: The ratio of the number of pink marbles to purple marbles in a bag is 17:13. If 24 pink marbles are removed, the new ratio of pink to purple is 1:1. Find the number of pink marbles and purple marbles before they were removed from the bag.

Solution: The difference in the number of parts for pink and purple marbles is 17 – 13 = 4. Thus, in order for the new ratio of pink to purple to be 1:1, the difference in the two color marbles must be 24 in order for 24 pink to be removed.

24 ÷ 4 = 6, which is the factor that must be multiplied to the ratio.

$$\text{Pink} = 17 * 6 = 102$$
$$\text{Purple} = 13 * 6 = 78$$

Check: Pink – Purple = 102 – 78 = 24

Answer: 102 pink marbles and 78 purple marbles

Review Problems

1. The odds of Jack losing to Jill at a game is 13 to 7. What is the probability that Jack will win? Express your answer as a percent.

2. A bag has red and green marbles and the ratio of red to green is 3 to 8. If there are 33 red marbles, how many green marbles are there?

3. There are 40 people in a room where the ratio of men to women is 3:5. What is the positive difference between the number of men and number of women in the room?

4. A bag has yellow and purple marbles with a ratio of 1 to 1. There is a total of 18 marbles. If 3 more yellow marbles were added, what is the new ratio of yellow to purple marbles?

5. The ratio of boys to girls in a fifth-grade class is 9 to 8. If there are 72 girls, how many total fifth-graders are there?

6. There are apples, oranges and bananas in a large basket. If the ratio of apples to oranges to bananas is 2:5:3 and the total number of fruits is 170, how many apples are there?

7. The odds that Daniel draws a red sock from a drawer with red, yellow and blue socks is 7 to 10. If it is equally likely to draw a yellow sock as it is to draw a blue sock, what fraction of the total number of socks is blue?

8. Jacob has nickels, dimes and quarters. The ratio of nickels to dimes to quarters is 11 to 7 to 6. If he has 72 total in nickels and dimes, how many quarters does he have?

9. The ratio of the side lengths of a quadrilateral is 8:6:4:6. If the perimeter is 264, find the positive difference between the longest and shortest side lengths.

10. In a parking lot, the ratio of the number of cars to the number of trucks is 1 to 2. If there are 99 more trucks than cars, how many total vehicles are in the parking lot?

11. The ratio of unicycles to bicycles to tricycles in a park is 3:2:1. If there are 20 more unicycles than tricycles, how many bicycles are there?

12. The ratio of DVD's to books on a shelf is 7 to 5. If there are 60 books, how many DVD's need to be removed from the shelf to give a new ratio of DVD's to books to be 2 to 3?

13. A room has pink and green balls. The ratio of pink to green balls is 29 to 14. If 195 pink balls were removed from the room, the new ratio would be 1 to 1. How many green balls are in the room?

*14. The ratio of boys to girls in a class is 5 to 4. If the number of girls were doubled, there would be 27 more girls than boys. What is the sum of the the number of boys and girls before the girls were doubled?

15. In a quadrilateral, the ratio of the angles is 4:5:11:4. Find the measure of the largest angle. (Hint: The sum of the angles for a quadrilateral is 360 degrees.)

16. There are squares, triangles and circles drawn on the driveway. The ratio of the number of squares to the number of triangles to the number of circles is 14:9:10. If there are 12 more squares than circles, how many total shapes are drawn on the driveway?

17. If the ratio of teachers to boys to girls is 2 to 8 to 7, what fraction of the school population is students?

18. The ratio of dogs to rabbits to cats at an animal shelter is 19 to 15 to 17. If 32 dogs and 16 cats are adopted, the new ratio of dogs to rabbits to cats would be 1 to 1 to 1. How many total animals are at the shelter after the adoptions?

19. If the ratio of teachers to boys to girls is 3 to 11 to 15 and there are 492 more girls than teachers, how many boys are there?

*20. A bag has orange and green marbles with a ratio of orange to green equal to 9 to 5. If 71 orange marbles are removed, there would still be 5 more orange marbles than green marbles. How many marbles are there before the orange marbles were removed from the bag?

Arithmetic Sequences

An **arithmetic sequence** is a sequence of numbers where the difference between consecutive terms is the same value. For example, the following is an arithmetic sequence where 1 is the first term: 1, 5, 9, 13, 17, ...

The **common difference** is the difference between any two consecutive terms. In this case, 5 – 1 = 4 or 9 – 5 = 4 which means that 4 is the common difference.

*n*th Term of an Arithmetic Sequence

To find the *n*th term of an arithmetic sequence, the following formula can be used:

$$a_n = a_1 + (n - 1)d$$

where a_n is the *n*th term, a_1 is the first term and d is the common difference.
$d = a_2 - a_1 = a_3 - a_2 = \ldots$

In the example above, to find the 5th term of the sequence, plug values into the formula:
$a_1 = 1, n = 5, d = 4$
$a_5 = 1 + (5 - 1)4 = 1 + 16 = 17$

Thus, the fifth term is 17 (which can also be seen in the sequence).

Sum of an Arithmetic Sequence

To find the sum of consecutive terms in an arithmetic sequence, the following formula can be used:

$$S_n = \frac{n(a_1 + a_n)}{2}$$

where S_n is the sum of the *n* terms, a_1 is the first term of the sequence and a_n is the *n*th term of the sequence.

Using the example above, to find the sum of the terms from the second term to the fifth term, plug values into the formula.

In this case, $a_1 = 5$ because it is the first term of the sequence to be summed. $n = 4$ because there are 4 terms. $a_n = 17$ which is the final term.

Therefore,

$$S = \frac{4(5+17)}{2} = \frac{88}{2} = 44$$

Example 1: In the arithmetic sequence, 11, 8, 5, 2, ... , what is the 10th term?

Solution:
$n = 10, a_1 = 11, d = 8 - 11 = -3$
$a_{10} = a_1 + (n-1)d = 11 + (10-1) * (-3) = 11 + 9 * (-3) = 11 - 27 = -16$

Answer: −16

Example 2: In the arithmetic sequence, 4, 10, 16, 22, 28, ... , what is the 101st term?

Solution:
$n = 101, a_1 = 4, d = 10 - 4 = 6$
$a_{101} = 4 + (101 - 1)6 = 4 + 100 * 6 = 604$

Answer: 604

Example 3: Find the missing terms in the arithmetic sequence: 87, ___, 121, 138, ___

Solution: The common difference can be found from 121 and 138: $d = 138 - 121 = 17$

To find the missing terms:
$a_2 = 87 + 17 = 104$
$a_5 = 138 + 17 = 155$

Answer: 104 and 155

Example 4: In the arithmetic sequence, 4, 9, 14, 19, …, what is the sum of the first 10 terms?

Solution:
First, find the 10th term.
$a_1 = 4, d = 9 - 4 = 5$
$a_{10} = 4 + (10 - 1)5 = 49$

Now, use the formula to calculate the sum of the terms.
$$S_{10} = \frac{10(4 + 49)}{2} = \frac{530}{2} = 265$$

Answer: 265

Example 5: Find the sum of the first 20 positive odd integers.

Solution: The sequence is 1, 3, 5, 7…
$a_1 = 1, d = 3 - 1 = 2$
$a_{20} = 1 + (20 - 1)2 = 39$

$$S_{10} = \frac{20(1 + 39)}{2} = \frac{800}{2} = 400$$

Answer: 400

Review Problems

1. Given an arithmetic sequence, 2, 5, 8, 11, …, find the 9th term.

2. Given an arithmetic sequence, −20, −15, −10, −5, …, find the 71st term.

3. Find the sum of the first 10 terms of the sequence: 2, 7, 12, 17, …

4. Find the sum of the first 15 terms of the sequence: −14, −12, −10, −8, …

5. Find the missing terms in the sequence: ___, 11, ___, 29, 38, ___

6. Find the missing terms in the sequence: -55, ___, 215, ___, ___, 620, 755

7. What is the next term in the sequence? 80, 71, 62, 53,…

8. If James' allowance is $15 dollars in January, $35 in February, $55 in March and so on with every monthly allowance increasing by $20, how much money would he have received in allowance in a year?

9. Find the sum of the first 15 positive integers.

10. Find the missing terms in the sequence: 88, ___, 112, ___, 136, ___

11. If a grandfather clock chimes once at 1 o'clock, twice at 2 o'clock, thrice at 3 o'clock and so on for each of the 12 hours, how many times does it chime in total over the 12 hour period?

12. What is the product of the 10th term and the 100th term of the following sequence: −297, −294, −291, … ?

13. Find the missing terms, then find the sum of the first 20 terms for the following sequence: 9, ___, 25, 33, ____, …

14. What is the sum of the terms between the 10th term and the 30th term, inclusive, of the following sequence: 1, 16, 31, 46, … ?

15. Find the sum of the first 25 positive even integers.

16. Find the positive difference between the sum of the first 30 positive even integers and the sum of the first 30 positive odd integers.

17. A store has a stack of soda cans. The top row has 1 can, the second row has 8 cans, and so on in an arithmetic sequence until the final row with 50 cans. How many cans in total are in the stack?

18. Find the missing terms: 11, ___, ___, ___, 251, ___

*19. Find the difference between the sum of the 1st and 50th term and the sum of the 2nd and 49th term of the sequence: $-146, -96, -46, \ldots$

*20. If the first term of a sequence is 3, the last term is 39 and the sum of the terms from first to last is 210, what is the product of the common difference and the number of terms?

Measurements

Units of Measurements

Length or Distance Conversions
1 foot = 12 inches
1 yard = 3 feet
1 mile = 5280 feet
1 meter = 100 centimeters
1 kilometer = 1000 meters

Time Conversions
1 minute = 60 seconds
1 hour = 60 minutes
1 day = 24 hours
1 non leap year = 365 days
1 leap year = 366 days

Volume Conversions
1 cubic foot = 1728 cubic inches
1 cubic yard = 27 cubic feet
1 liter = 1,000 cubic centimeters

Capacity Conversion
1 cup = 8 fluid ounces
1 pint = 2 cups
1 quart = 2 pints
1 gallon = 4 quarts
1 liter = 1000 milliliters

Area Conversion
1 square foot = 144 square inches
1 square yard = 9 square feet
1 square mile = 640 acres

Weight and Mass Conversion
1 pound = 16 ounces
1 kilogram = 1000 grams
1 ton = 2,000 pounds

Factor Label Method

When converting from one unit to another, the factor label method can be used to keep track of the units. In this method, conversion factors, which can be found above, are set up as fractions that are essentially equal to 1. The conversion factors are multiplied so the units are on the top and on the bottom, thus cancelling units.

Follow these steps when using the Factor Label Method

Step 1) Set up the first fraction
Step 2) Line up conversion factors to alternate units
Step 3) Cancel the units
Step 4) Multiply the numbers and keep the remaining unit(s)

Example 1: How many inches are in 4 feet?

Solution:

1) Set up the first fraction: $\frac{4 \text{ feet}}{1}$

2) Line up conversion factors: $\frac{4 \text{ feet}}{1} * \frac{12 \text{ inches}}{1 \text{ feet}}$ (1 foot = 12 inches)

3) Cancel the units: $\frac{4}{1} * \frac{12 \text{ inches}}{1}$

4) Multiply the numbers and keep the remaining unit: 48 inches

Answer: 48 inches

Example 2: Convert 2 yards to inches.

Solution:

1) Set up the first fraction: $\frac{2 \text{ yards}}{1}$

2) Line up conversion factors: $\frac{2 \text{ yards}}{1} * \frac{3 \text{ feet}}{1 \text{ yards}} * \frac{12 \text{ inches}}{1 \text{ feet}}$

3) Cancel the units: $\frac{2}{1} * \frac{3}{1} * \frac{12 \text{ inches}}{1}$

4) Multiply the numbers and keep the remaining unit: 72 inches

Answer: 72 inches

Example 3: How many square inches are in a square foot?

Solution:

1) Set up the first fraction: $\frac{1 \text{ feet}^2}{1}$

2) Line up conversion factors: $\frac{1 \text{ feet}^2}{1} * \frac{12 \text{ inches}}{1 \text{ feet}} * \frac{12 \text{ inches}}{1 \text{ feet}}$

3) Cancel the units: $\frac{1}{1} * \frac{12 \text{ inches}}{1} * \frac{12 \text{ inches}}{1}$

4) Multiply the numbers and keep the remaining unit: 144 inches^2

Answer: 144 inches^2

Example 4: How many cubic feet are in a cubic yard?

Solution:

1) Set up the first fraction: $\frac{1 \text{ yard}^3}{1}$

2) Line up the conversion factors: $\frac{1 \text{ yard}^3}{1} * \frac{3 \text{ feet}}{1 \text{ yard}} * \frac{3 \text{ feet}}{1 \text{ yard}} * \frac{3 \text{ feet}}{1 \text{ yard}}$

3) Cancel the units: $\frac{1}{1} * \frac{3 \text{ feet}}{1} * \frac{3 \text{ feet}}{1} * \frac{3 \text{ feet}}{1}$

4) Multiply the numbers and keep the remaining unit: 27 feet^3

Answer: 27 feet^3

Example 5: Angelina drives from her house to work at an average speed of 45 miles per hour. What is her speed in feet per second?

Solution:

1) Set up the first fraction: $\frac{45 \text{ miles}}{1 \text{ hours}}$

2) Line up the conversion factors: $\frac{45 \text{ miles}}{1 \text{ hours}} * \frac{1 \text{ hours}}{60 \text{ minutes}} * \frac{1 \text{ minutes}}{60 \text{ seconds}} * \frac{5280 \text{ feet}}{1 \text{ mile}}$

3) Cancel the units: $\frac{45}{1} * \frac{1}{60} * \frac{1}{60 \text{ seconds}} * \frac{5280 \text{ feet}}{1}$

4) Multiply the numbers and keep the remaining units: $\frac{66 \text{ feet}}{1 \text{ second}}$

Answer: 66 feet per second

Chapter 11

Review Problems

1. Convert 5 pounds to ounces.

2. How many hours are in 7 days?

3. How many yards are in a mile?

4. How many seconds are in 30 minutes?

5. What is 35 feet in inches?

6. An action-figure weighs 8 pounds and 4 ounces. How much does it weigh in ounces?

7. How many fluid ounces are in 4 gallons of milk?

8. If Joe needs 90 square feet of tile to complete his floor, how many square yards does he need to buy?

9. If there are 12 jegs in a jag, 4 jags in a jug, and 10 jugs in a jog, how many jegs are in a jog?

10. Pauline rides her bike at an average speed of 15 feet per second. What is her speed in miles per hour? Express your answer as a reduced improper fraction.

11. If there are 3 kones in a kylinder, 4 kylinders in a kube, and 5 kubes in a kettle, how many kones are in a kone, a kylinder, a kube and a kettle?

12. Alan's family likes to drink water. If they consume 25 gallons, 13 quarts, and 16 pints of water in a week, how many quarts of water did the family consume?

13. If a palm tree at the beach stands 2 yards, 4 feet and 5 inches tall, how tall is the tree in inches?

14. An action movie's feature-length is 2 hours, 34 minutes and 15 seconds long. How long is the movie in seconds?

15. If $1.80 = 1 Mars dollar, how much does an item priced at $8.10 cost in Mars dollars?

16. How many cubic inches are in ⅜ of a cubic yard?

17. The mail carrier drives his truck for 840 miles every week. If he works 8 hours per day from Monday through Friday, what is his average driving speed in miles per hour?

18. A machine lifts 9 tons of building materials per hour. How much does it lift in ounces per second?

19. Anna is driving at a speed of 64 miles per hour while Kyle is driving at a speed of 49 miles per hour. How much faster is Anna driving in feet per second?

20. Darry mows grass at an average rate of 660 square feet per hour. What is his average speed in square inches per minute?

Chapter 12

Distance and Work Formula

Distance = Rate × Time

D = R × T

The distance traveled is equal to the rate or the speed multiplied by the time.

Work = People × Rate × Time

W = P × R × T

In the United States, speed is expressed as miles per hour (mph). It indicates the number of miles that would be traveled in one hour if that speed was kept constant.

20 miles per hour can be written as: $\frac{\text{20 miles}}{\text{1 hour}}$

Example 1: If a baseball travels at a constant speed of 100 mph for 12 seconds, how far does the baseball travel? Express your answer in miles.

Solution: Using D = R × T, R = 100 mph and T = 12 seconds.
T must be converted into hours.

$$\frac{\text{12 seconds}}{1} * \frac{\text{1 minute}}{\text{60 seconds}} * \frac{\text{1 hour}}{\text{60 minutes}} = \frac{1}{300} \text{ hour}$$

Now, the distance formula can be used to calculate the distance traveled.

$$D = R * T = \frac{\text{100 miles}}{\text{1 hour}} * \frac{\text{1 hour}}{300} = \frac{1}{3} \text{ mile}$$

Answer: ⅓ mile

Example 2: Josh can make 10 paper airplanes every 5 minutes. How long would it take him to produce 114 paper airplanes, assuming he continues to make planes at a constant rate? Express your answer in minutes.

Solution: Using W = P × R × T, P = 1 person, W = 10 planes and T = 5 minutes.
Since there is one person, W = R × T

R = W ÷ T = 10 ÷ 5 = 2 planes per minute or $\frac{2 \text{ planes}}{1 \text{ minute}}$

W = 114 planes

T = W ÷ R = 114 planes ÷ (2 planes per minute) = $114 \text{ planes} * \frac{1 \text{ minute}}{2 \text{ planes}} = 57 \text{ minutes}$

Answer: 57 minutes

Example 3: Robert went on a three-day vacation. On the first day, he drove for 6 hours and traveled 260 miles. On the second day, he drove for 11 hours and traveled 640 miles. And on the last day, he drove for 7 hours and traveled 300 miles. What was his average speed over the three days? Express your answer in miles per hour.

Solution: To find the average speed, the total distance and the total travel time must be calculated

D = 260 + 640 + 300 = 1200 miles
T = 6 + 11 + 7 = 24 hours

Using D = R × T,

R = D ÷ T = 1200 miles ÷ 24 hours = 50 miles per hour

Answer: 50 miles per hour

Example 4: Janet paints at a rate of 10 square yard per hour. If the surface area of the wall is 810 square feet, how many minutes will it take Janet to paint the entire wall?

Solution: R = 10 square yard per hour
This must be converted to square feet per minute to solve the problem.

$$R = \frac{10 \text{ yard}^2}{1 \text{ hour}} * \frac{3 \text{ feet}}{1 \text{ yard}} * \frac{3 \text{ feet}}{1 \text{ yard}} * \frac{1 \text{ hour}}{60 \text{ minutes}} = \frac{3 \text{ feet}^2}{2 \text{ minutes}}$$

R = 3/2 square feet per minute
W = 810 square feet

Using W = R × T,
T = W ÷ R = 810 square feet ÷ 3/2 square feet per minute =

$$\frac{810 \text{ feet}^2}{1} * \frac{2 \text{ minutes}}{3 \text{ feet}^2} = 540 \text{ minutes}$$

Answer: 540 minutes

Example 5: If three factory workers can paint 120 dolls in 5 days, how many days would it take for six workers to paint 960 dolls?

Solution:
Dolls = work done
Workers = people
Days = time
Units are the same (dolls, workers and days) so no conversions are necessary.

W = P × R × T, W = 120, P = 3, T = 5

W = P × R × T
120 = 3 × R × 5 = R × 15
R = 120 ÷ 15 = 8

Now, the question asks for T where P = 6 and W = 960.

960 = 6 × 8 × T = T × 48
T = 960 ÷ 48 = 20

Answer: 20 days

Review Problems

1. Joey drove his car at a constant speed of 40 miles per hour for 30 minutes. How far did he travel? Express your answer in miles.

2. Bob drives 24 miles from his house to work. If it takes him 40 minutes to arrive at his destination, what is his average driving speed in miles per hour?

3. Leo works at a toy factory. If he makes toys at a constant rate of 12 toys every hour, how many toys would he make in 8 hours?

4. Jarvis bikes at a constant rate and traveled 5 miles in 20 minutes. How many hours would it take him to bike 45 miles?

5. Kyle sewed 525 shirts in 75 hours. What was his average rate of shirts sewed per hour?

6. A plane travels a distance of 1125 miles at an average speed of 540 miles per hour. How many minutes does the trip take?

7. During a road trip, Jonathan drives at a rate of 45 miles per hour for the first two hours, a rate of 60 miles per hour for the next five hours and a rate of 30 miles per hour for the final three hours. What is his average speed in miles per hour for the whole road trip?

8. Carl can solve 14 math problems per hour. If his older brother can solve twice as many math problems per hour as Carl, how many hours will it take them to solve 336 math problems if they work together?

9. Three workers can build 72 microwaves in 8 hours. How many microwaves can one worker build in one hour? Hint: It is equivalent to the rate.

10. Julian lives 120 miles from his mom. If he travels at an average speed of 40 miles per hour going to his mom's house and 24 miles per hour on the return trip, what is his average speed in miles per hour for the entire trip?

11. 6 workers can build 378 chairs in 7 hours. How many chairs can 4 workers build in 5 hours?

12. A giant water fountain dispenses 40 quarts of water every minute. How many hours would it take to dispense 960 gallons of water? Express your answer as a mixed number.

13. James has three lawns where each lawn has 1800 square feet of area. If he can mow at a constant rate of 720 square inches per minute, how many hours will it take him to finishing mowing his three lawns?

14. 12 workers can make 1440 DVD's in 15 hours. If each worker works at the same constant rate, 21 workers can make 840 DVD's in X minutes. What is X?

15. Angela can feed 7 cows every hour, Bessie can feed 8 cows every hour and Cassie can feed 10 cows every hour. If they all work together, how many hours would it take them to feed 300 cows?

16. Robin is driving ahead of Will by 20 miles traveling in the same direction. If Robin is driving at a constant speed of 39 miles per hour and Will is driving at a constant speed of 51 miles per hour, how many minutes will it take Will to catch up to Robin?

17. Wally can make 6 pens in one hour. Jay can make 10 pens in half an hour. Barry can make 12 pens in 45 minutes. How many pens can they make together in 2 hours?

18. Gerald can print 2 shirts every 5 minutes. Identical twins Lucy and Lucas can each print 7 shirts every 10 minutes. How many minutes would it take for the three of them working together to print 378 shirts?

19. Abby can make a button every 10 seconds, Beatrice can make a button every 15 seconds and Cassandra can make a button every 30 seconds. If they work together, how many minutes would it take them to make 408 buttons?

20. Jack and Jill are 1080 miles apart. They drive at different speeds toward each other to meet up. If Jack drives at an average speed of 37 mph and Jill drives at an average speed of 53 mph, how many hours will it take them to meet up?

Fundamental Theorem of Counting

If there are x ways of an event occurring and y ways of another event occurring, the number of total possible outcomes is $x \times y$. This is known as the Fundamental Theorem of Counting.

Useful Facts:
There are 10 digits (0 to 9).
There are 26 letters in the alphabet (A to Z).
There are 5 vowels in the alphabet (A, E, I, O, U).
There are 21 consonants (non-vowels) in the alphabet (26 – 5 = 21).
There are 2 sides of a coin (head and tail)
There are 6 sides or faces of a die.

Example 1: Billy wants to create his own ice cream cone. He can choose from 3 types of cones and 2 types of flavors. How many different combinations can he make if he chooses one from each category?

Solution: By the Fundamental Theorem of Counting, multiply 3 and 2.

$3 \times 2 = 6$ different combinations

This can also be viewed visually:

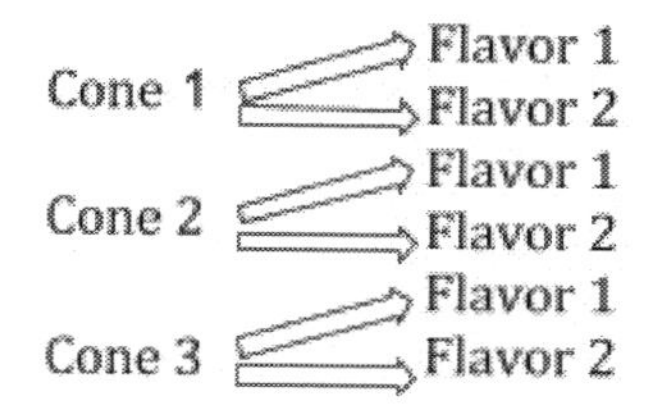

Each cone branch has 2 flavors and since there are 3 cone branches, the product is the total number of combinations ($3 \times 2 = 6$)

Answer: **6 combinations**

Example 2: How many different ways can a coin be flipped and a die be rolled?

Solution: A coin has 2 sides and a die has 6 faces. Now, calculate the product.

$2 \times 6 = 12$

Answer: 12 ways

Example 3: How many different ways can three coins be flipped?

Solution: Each coin has 2 outcomes.

$2 \times 2 \times 2 = 8$

Answer: 8 ways

Example 4: How many combinations of 2 letters followed by a digit are there?

Solution: Each letter has 26 options and a digit has 10 options. Now, calculate the product.

$26 \times 26 \times 10 = 6{,}760$

Answer: 6,760 combinations

Example 5: How many four-digit numbers are there? (Note: 0134 is not a four-digit number.)

Solution: The leading digit cannot be a 0 or else the number would not be a four-digit number. Thus, there are 9 options for the first digit (1 - 9). The last three digits each have 10 options (0 - 9). Now, calculate the product.

$9 \times 10 \times 10 \times 10 = 9{,}000$

Answer: 9,000 numbers

Review Problems

1. How many two letter “words” consisting of only vowels are there?

2. How many two letter “words” consisting of only consonants are there?

3. How many ways can three dice of different colors be rolled?

4. How many ways can a blue die and a green die be rolled and a quarter be flipped?

5. How many different license plates consisting of one consonant followed by two digits are there?

6. How many two letter "words" consisting of a consonant followed by a vowel are there?

7. How many five-digit numbers are there?

8. What is the sum of the numbers of three-digit, two-digit and one-digit numbers? (Note: 0 is a one-digit number)

9. Robert wants to buy a sandwich. He can choose from 3 types of bread, 5 types of meat, 4 types of vegetables and 2 types of cheese. If he picks one from each category, how many different sandwich combinations can he make?

*10. Jackie wants to buy a sandwich. She can choose from 3 types of bread, 4 types of meat, 3 types of vegetables and 9 types of cheese. If she picks one from each category with the exception that she can omit cheese, how many different sandwich combinations can she make?

11. How many three letter "words" are there if the letters cannot be repeated?

12. How many two-digit numbers are there if no digits can be repeated?

13. How many four letter "words" are there if the first letter is the same as the last letter?

*14. How many three-digit numbers are there if consecutive pairs of digits are not the same digit? (i.e. 122 has a consecutive pair of "22")

*15. How many two letter "words" consisting of a consonant and a vowel are there? Hint: Order matters.

16. How many five-digit even numbers can be created?

*17. Max wants to buy a sandwich. He can choose from 5 types of bread, 7 types of meat, 6 types of vegetables and 3 types of cheese. If he is allowed to pick one from each category with the option of omitting meat, vegetable and/or cheese, how many different sandwich combinations can he make?

18. How many four-digit numbers can be created such that the number is divisible by 10 and no digit is repeated?

19. How many license plates with three-letters followed by three-digits can be created if the first two letters are vowels, the first two digits are divisible by 3 and the last digit is divisible by 5?

*20. How many four letter "words" are there if consecutive pairs of letters cannot be the same letter? (i.e. book has a consecutive pair of "oo")

Factorials, Permutations, Combinations

The factorial of a whole number, *n*, is the product of all the whole numbers less than or equal to *n*. We use *n* factorial to find the number of ways to arrange *n* items in a line.
$n! = n * (n-1) * \ldots * 3 * 2 * 1$

It can also be written: $n! = n * (n-1)!$

Factorials

$0! = 1$
$1! = 1$
$2! = 2 * 1 = 2$
$3! = 3 * 2 * 1 = 6$
$4! = 4 * 3 * 2 * 1 = 24$
$5! = 5 * 4 * 3 * 2 * 1 = 120$
$6! = 6 * 5 * 4 * 3 * 2 * 1 = 720$
$7! = 7 * 6 * 5 * 4 * 3 * 2 * 1 = 5040$

To calculate the number of ways to arrange items in a circle, we can divide $n!$ by *n* to get: $(n-1)!$ ways to arrange items in a circle, since a circle can be rotated *n* times for *n* items.

Permutations (Order Matters)

For example, if there are 5 different books, where Arnold and Bobby are given one book each, how many different ways can the books be distributed?

Starting with Arnold, he has 5 choices for his book. This leaves 4 choices for Bobby. Multiplying the two numbers: $5 * 4 = 20$ (ways)

Permutation Formula:

$$P(n, r) = {}_nP_r = \frac{n!}{(n-r)!}$$

where *n* is the number of items and *r* is the number of items being distributed. Use this formula when the order of the items are drawn matters. In this example, Arnold and Bobby are different people receiving different books so order matters.

$$P(5, 2) = {}_5P_2 = \frac{5!}{(5-2)!} = \frac{5!}{3!} = \frac{120}{6} = 20$$

Combinations (Order Does Not Matter)

For example, if there are 5 different books, and Cathy selects two books from them, how many different ways can the books be chosen?

This seems like it would be solved the same way as with the Permutations problem, however, the order does not matter in this problem: (i.e. choosing Book A and then Book B is the same as choosing Book B and then Book A.)

Thus, there are 5 ways to choose the first book and 4 ways to choose the second book but the product must be divided by $2!$ which is the number of ways the two books can be rearranged.

$$\frac{5*4}{2!}=\frac{20}{2}=10$$

Combination Formula:

$$C(n,r)={}_{n}C_{r}=\frac{n!}{r!(n-r)!}$$

where *n* is the number of items and *r* is the number of items being distributed. Use this formula when the order of the items are drawn does not matter. In this example, the order that Cathy draws the books does not matter.

$$C(5,2)={}_{5}C_{2}=\frac{5!}{2!(5-2)!}=\frac{120}{2*6}=10$$

Example 1: How many ways can 3 people stand in a line?

Solution: For the first spot of the line, there are 3 people to choose from. This leaves 2 people to choose from to fill the second spot. Finally, this leaves 1 person for the final spot. Now, multiply these numbers. $3*2*1=6$. Note: This is the same as $3!=6$.

Answer: **6 ways**

Example 2: How many ways can 5 children be seated in a circle?

Solution: Using the formula, there are $(n-1)!$ ways to arrange *n* items in a circle. Thus, $(5-1)!=4!=24$.

Answer: **24 ways**

Example 3: 10 people are running for student council positions. The winner will be the president and the runner-up is the vice president. How many different outcomes can there be?

Solution: This is a Permutations problem because order matters (the president is different from the vice-president). There are 10 people and 2 people to be chosen.

$$_{10}P_2 = \frac{10!}{(10-2)!} = \frac{10*9*8!}{8!} = 10*9 = 90$$

Answer: **90 outcomes**

Example 4: Tommy has 8 different toy cars. If he takes 3 of these toy cars with him when he leaves the house, how many different sets of 3 toy cars can he bring?

Solution: This is a Combinations problem because the order does not matter (the 3 cars can be chosen in any order). There are 8 items and 3 items to be chosen.

$$_{8}C_3 = \frac{8!}{3!*(8-3)!} = \frac{8*7*6*5!}{3!*5!} = \frac{8*7*6}{3!} = \frac{8*7*6}{6} = 56$$

Answer: **56 sets**

Example 5: How many ways can the letters in the word ARRAY be arranged?

Solution: Examining the word ARRAY, there are 2 As, 2 Rs and 1 Y. If all the letters were different, the answer would be $5!$ but there are 2 As and 2 Rs. Thus, we must divide by the number of ways that the As, Rs and Y can be rearranged within its own letter. The 2 As can be rearranged $2!$ ways, the 2 Rs can be rearranged $2!$ ways and the 1 Y can be rearranged $1!$ ways.

$$\frac{5!}{2!*2!*1!} = \frac{120}{4} = 30$$

Answer: **30 ways**

Chapter 14

Review Problems

1. How many ways can the digits in the number 1,234 be rearranged?

2. How many ways can 6 people finish a race given that there are no ties?

3. How many ways can 8 different colored beads be placed in a circle on a table?

*4. A bead bracelet consists of 7 different colored beads. How many different ways can the beads be arranged? Hint: The bracelet can be flipped.

5. How many ways can 10 people enter a building if one person goes in at a time?

6. 11 people are running for office. There are three positions available: president, secretary and treasurer. One person can only hold one position. How many ways can these positions be filled up?

7. 8 people are running in a race. How many ways can the top 3 runners finish? (Assuming there are no ties)

8. Josie has 6 favorite pens of different colors. If she can only take 3 to school, how many different sets of pens can she bring?

9. John has 10 wood blocks of different shapes. How many ways can he choose 8 blocks?

10. A bag has 5 different colored marbles: red, orange, green, blue, and yellow. How many ways can 3 of the marbles be drawn from the bag if they are all drawn at the same time?

11. A company wants to choose a group of 4 representatives from 10 employees to send to a conference. How many ways can this group be formed?

*12. There are 4 boys and 4 girls standing in a line waiting to purchase movie tickets. How many ways can they stand in a line if all the boys must stand together and all the girls must stand together?

13. James has 10 favorite toys. He needs to choose 2 of them for one bag and then choose 3 of them for one box. How many ways can he choose these 5 toys to store away?

14. There are 6 balls arranged in a row: 2 red, 1 green, 1 blue, 1 yellow and 1 purple. If the red balls are identical, how many ways can the balls be arranged?

15. There are 6 balls arranged in a row: 2 red, 2 white, 1 blue and 1 yellow. If the 2 red balls are identical and the 2 white balls are identical, how many ways can the balls be arranged?

16. How many ways can the letters in the word CREAM be arranged?

17. How many ways can the digits in the number 12,556 be arranged?

18. How many ways can the letters in the word RACECAR be arranged?

19. How many ways can the digits in the number 7,844,847 be arranged?

20. How many ways can the letters in the word AARDVARK be arranged?

Statistics: Mean, Median, Mode and Range

Given a set of numbers, we can find the mean, median, mode and range.

Mean = Average value

Mean = $\frac{\text{(Sum of All Terms)}}{\text{(Number of Terms)}}$

Thus, $\text{(Sum of All Terms)} = \text{(Number of Terms)} * \text{(Mean)}$

Median = Middle value
In a set containing an odd number of terms, the median is the middle number.
In a set containing an even number of terms, the median is the average of the two middle numbers.

Mode = The value that appears the most
Note: There may be more than one mode. If all of the numbers are different, there is no mode.

Range = The difference between the largest number and the smallest number
Range = Largest Value – Smallest Value

Steps to solve: Mean, Median, Mode and Range Problems

1) Arrange the numbers from least to greatest
2) Count the number of terms
3) Calculate the Mean, Median, Mode and Range

Example 1: Given a set of numbers, {22, 42, 14}, find the mean, median, mode and range.

Solution:
1) Arrange the numbers: {14, 22, 42}
2) There are 3 terms
3)

Mean: $\frac{14+22+42}{3} = \frac{78}{3} = 26$

Median: There is an odd number of terms so 22 is the middle number or the median

Mode: No mode because no value appears more than once.

Range: 42 – 14 = 28

Answer:
Mean: 26
Median: 22
Mode: None
Range: 28

Example 2: Given a set, {12, 19, 12, 17, 17, 19}, find the mean, median, mode and range.

Solution:
1) Arrange the numbers: {12, 12, 17, 17, 19, 19}
2) There are 6 terms
3)

Mean: $\frac{12+12+17+17+19+19}{6} = \frac{96}{6} = 16$

Median: There is an even number of terms so the median is the average between the two middle numbers. $\frac{17+17}{2} = \frac{34}{2} = 17$

Mode: 12, 17, 19 all appear more than once at the same frequency (2 times). Therefore, all three are modes.

Range: 19 – 12 = 7

Answer:
Mean: 16
Median: 17
Mode: 12, 17, 19
Range: 7

Example 3: Jack likes to play on his pinball machine and keep track of his scores. If he scored 300, 250, 275, 250, and 225 points on his past five games, what is the mean, median, mode and range of his scores?

Solution:
1) Arrange the numbers: {225, 250, 250, 275, 300}
2) There are 5 terms
3)

Mean: $\frac{225 + 250 + 250 + 275 + 300}{5} = \frac{1300}{5} = 260$
Median: 250
Mode: 250
Range: 300 – 225 = 75

Answer:
Mean: 260 points
Median: 250 points
Mode: 250 points
Range: 75 points

Example 4: Janet has a garden consisting of 8 plants with heights: 135 cm, 120 cm, 58 cm, 46 cm, 99 cm, 78 cm, 105 cm, and 110 cm. What is the positive difference between the median of the 8 plants' heights and the tallest plant's height?

Solution:
1) Arrange the numbers: {46, 58, 78, 99, 105, 110, 120, 135}
2) There are 8 plants
3)

Median: $\frac{99 + 105}{2} = \frac{204}{2} = 102$
Tallest: 135

$135 - 102 = 33$

Answer: **33 cm**

Example 5: Given a set, { 54, 18, 34, 22, -10, 20, 90, 17, 34, 61}, let A be the mean, B be the median, C be the mode and D be the range. Find the mean of the set: {A, B, C, D}

Solution:

1) Arrange the numbers: {-10, 17, 18, 20, 22, 34, 34, 54, 61, 90}
2) There are 10 terms
3)

A = Mean: $\frac{-10+17+18+20+22+34+34+54+61+90}{10} = \frac{340}{10} = 34$

B = Median: $\frac{22+34}{2} = 28$

C = Mode: 34

D = Range: $90-(-10)$ = 100

{A, B, C, D}: {34, 28, 34, 100} (4 terms)

Mean = $\frac{34+28+34+100}{4} = \frac{196}{4} = 49$

Answer: **49**

Review Problems

1. Given a set, {17, 33, 5, 25}, find the median.

2. Given a set, {22, 28, 29, 22, 34}, find the mean.

3. Given a set, {31, 22, 58, 99, 79, 124}, find the range.

4. Given a set, {17, 44, 81, 44, 99, 17, 17, 81, 99, 81}, find the mode.

5. Given a set, {12, 16, 14, 14, 19, 18, 13, 14}, find the mean, median, mode and range.

6. What is the average angle of a triangle with angles: 108, 44, and 28 degrees?

7. A marathon runner ran distances: 25 miles, 13 miles, 17 miles, 12 miles and 18 miles. What was the average distance that the runner ran?

8. Over the course of 5 vacation days, Jake drove 23 miles, 43 miles, 37 miles, 14 miles and 13 miles. What is the median distance that he drove?

9. Lucy works at her lemonade stand. This past week, she sold $40, $33, $18, $33, $12, $18 and $18 of lemonades. What is the positive difference between the range and mode of her daily sales?

10. What is the average angle of a hexagon with angles: 120, 125, 120, 95, 136, and 124 degrees?

11. Given a set, {1, 11, 111, 1111, 11111, 111111}, find the positive difference between the median and the range.

12. Five movies have feature-lengths of 144 minutes, 126 minutes, 125 minutes, 131 minutes and 154 minutes. What is the sum of the mean and the range of the movies' feature-lengths?

13. Given a set, {34, 56, 71, 44, 27, 14}, find the positive difference between the mean and the range.

14. At the toy store, 8 different toys ranged in prices of $19, $45, $104, $99, $9, $12, $13 and $19. Let A be the mean, B be the mode and C be the median. Find the median of the set: {A, B, C}.

15. Given a set, {11, A}, the mean is 8. Find the value of A.

*16. Given a set, {33, 54, 20, 12, 19, 96}, let A be the mean, B be the smallest value and C be the largest value. Find the mean of set: {A, B, C}

17. Given a set, {A, B, 13, 15}, the mean is 12. Find the sum of A and B.

18. Three trees have heights of A, B, and 10 feet. If the mode of the trees' heights is 7 feet, what is the mean of the trees' heights?

19. Given an arranged set, {14, 16, A, 20, 22, B}, the median and the mean is 20. Find the positive difference between A and B.

*20. Given an arranged set, {10, 16, 17, A, B, 23, C}, the mean and median is 19 and the range is 18. Find the product of the median and range of set: {A, B, C}

Probability

Probability is the likeliness or chance that an event is going to occur and is measured by the ratio of favorable outcomes to the total number of possible outcomes. The probability of an event occurring is always between 0 and 1, inclusive.

$$\text{Probability} = \frac{\text{Favorable Outcomes}}{\text{Total Number of Possible Outcomes}}$$

The sum of all the probabilities that an event could or could not occur is always 1.

In other words, $\text{P(Event Occurring)} + \text{P(Event Not Occurring)} = 1$

Sometimes, finding the probability of an event not occurring and subtracting that probability from 1 is more efficient than calculating the probability of an event occurring. This is known as **complementary counting**.

Flipping a Coin

Flipping a coin has two outcomes: head or tail.

$$\text{P(head)} = \frac{1}{2} \qquad \text{P(tail)} = \frac{1}{2}$$

$$\text{P(head)} + \text{P(tail)} = \frac{1}{2} + \frac{1}{2} = 1$$

Rolling a Die

Rolling a die has six outcomes: 1, 2, 3, 4, 5, or 6.

$$\text{P}(1) = \frac{1}{6} \quad \text{P}(2) = \frac{1}{6} \quad \text{P}(3) = \frac{1}{6} \quad \text{P}(4) = \frac{1}{6} \quad \text{P}(5) = \frac{1}{6} \quad \text{P}(6) = \frac{1}{6}$$

$$\text{P}(1) + \text{P}(2) + \text{P}(3) + \text{P}(4) + \text{P}(5) + \text{P}(6) = \frac{1}{6} + \frac{1}{6} + \frac{1}{6} + \frac{1}{6} + \frac{1}{6} + \frac{1}{6} = 1$$

Example 1: A bag has 7 orange, 5 green, and 2 blue marbles. What is the probability of drawing an orange marble without looking?

Solution: There are 7 + 5 + 2 = 14 marbles in the bag and 7 of the marbles are orange.

Thus, the probability of drawing an orange marble is the ratio of the number of orange marbles to the total number of marbles.

$$\text{P(orange)} = \frac{7}{7+5+2} = \frac{7}{14} = \frac{1}{2}$$

Answer: $\frac{1}{2}$

Example 2: If a die is rolled, what is the probability that the number rolled is less than 5?

Solution: The six outcomes are 1, 2, 3, 4, 5 and 6. Four of these outcomes (1, 2, 3 and 4) are less than 5.

$$\text{P(Less than 5)} = \frac{4}{6} = \frac{2}{3}$$

Answer: $\frac{2}{3}$

Example 3: If two coins are flipped, what is the probability that they are both heads?

Solution: By the fundamental theorem of counting, there are four possible outcomes (2 × 2) for two coins being flipped.
These outcomes can be listed as follow:

Coin #1	Coin #2
Head	Head
Head	Tail
Tail	Head
Tail	Tail

There are four possible outcomes and only 1 of them are both heads.

$$\text{P(Both Heads)} = \frac{1}{4}$$

Another way to solve it is using the fundamental theorem of counting.

$$\text{P(Both Heads)} = P_1(\text{Head}) * P_2(\text{Head}) = \frac{1}{2} * \frac{1}{2} = \frac{1}{4}$$

Answer: $\frac{1}{4}$

Example 4: If three coins are flipped, what is the probability that there is at least one head?

Solution: By the fundamental theorem of counting, there are eight possible outcomes (2 × 2 × 2 = 8) for three coins being flipped.

There are four possible outcomes for how many heads show up (0, 1, 2 or 3). Thus, the question is asking for the probability that there is 1, 2 or 3 heads.

Instead of listing out each possibility or calculating the probability for each number of heads, complementary counting can be used.

Finding the probability that there are no heads and then subtracting that probability from 1 will give the probability that there is at least one head.

The fundamental theorem of counting can be used to calculate this probability:

$$\text{P(0 Heads)} = \text{P(No Head)} * \text{P(No Head)} * \text{P(No Head)} = \frac{1}{2} * \frac{1}{2} * \frac{1}{2} = \frac{1}{8}$$

Given: $\text{P(No Head)} = \text{P(Tail)} = \frac{1}{2}$

Complementary Counting: $\text{P(At Least 1 Head)} = 1 - \text{P(0 Heads)} = 1 - \frac{1}{8} = \frac{7}{8}$

Answer: $\frac{7}{8}$

Example 5: A bag contains 84 marbles where each marble is one of three colors, brown, pink and purple. ½ of the marbles are brown and ⅙ of the marbles are pink. If 16 additional purples marbles are added, what is the probability of drawing a purple marble?

Solution: To find the number of purple marbles, subtract the number of brown and pink marbles from the total number of marbles.

Brown = $84 * \frac{1}{2} = 42$

Pink = $84 * \frac{1}{6} = 14$

Purple = $84 - 42 - 14 = 28$

Since 16 additional purple marbles are added to the bag, the probability is:

$$\text{P(purple)} = \frac{28 + 16}{84 + 16} = \frac{44}{100} = \frac{11}{25}$$

Answer: $\frac{11}{25}$

Review Problems

1. If a die is rolled, what is the probability the number rolled is even?

2. If two coins are flipped, what is the probability that there is no tail?

3. A bag contains 5 blue, 11 green, 4 yellow and 2 orange marbles. If a marble is chosen at random, what is the probability that it is not a green marble?

4. A die is rolled 4 times. The results are 3, 3, 6 and 4. If the die is rolled again, what is the probability that it rolls a 2?

5. If a die is rolled and a coin is flipped, what is the probability the coin flips a head and the die rolls a 4?

6. If two dice are rolled, what is the probability that the sum of the numbers rolled is 1?

7. If a number is randomly chosen from 1 to 10, inclusive, what is the probability the number chosen is a prime number?

8. If two dice are rolled, what is the probability that the sum of the numbers rolled is 2?

9. A bag has 20 marbles in which ½ are blue, ⅕ are yellow and the rest are red. If 7 blue marbles, 5 yellow marbles and 6 red marbles are added to the bag, what is the probability that a blue marble is not drawn?

10. If two coins are flipped, what is the probability that there is at least one head?

11. What is the probability that a number chosen from 1 to 100, inclusive, is not divisible by 14?

12. A bookshelf holds 14 math books, 13 history books and 11 science books. If 2 books from each subject are removed from the shelf, what is the probability that a book chosen randomly from the shelf is a history book?

13. Jake has a bag of 29 gumballs. There are 5 red gumballs, three times as much green gumballs as red gumballs and the remaining gumballs are blue. If he randomly chooses one gumball from the bag, what is the probability that the gumballs is blue?

14. If three coins are flipped, what is the probability that there are two heads and one tail?

15. A coin is flipped and two dice are rolled. What is the probability the coin is not heads and the sum of the numbers rolled on the dice are 12?

*16. A bag has 10 marbles: 5 red, 3 yellow and 2 blue. What is the probability that a red marble is drawn first then a yellow marble is drawn second without replacement?

17. If four coins are flipped, what is the probability that there is at least one head?

18. What is the probability that a number chosen from 1 to 100, inclusive, is divisible by 3 but not divisible by 9?

*19. If three dice are rolled, what is the probability that either all 3 numbers are odd or all 3 numbers are even?

*20. What is the probability that a number chosen from 1 to 1000, inclusive, is divisible by 10 but not divisible by 3?

Geometric Figures I

This chapter will explore simple two dimensional (2D) shapes.

The following 2D shapes consist of sides and angles. We can calculate the **perimeter** by adding the lengths around a figure.

An angle, which is usually measured in degrees (°), is where two sides intersect. Two angles that add up to 90 degrees are called **complementary** angles. Two angles that add up to 180 degrees are called **supplementary** angles.

A **triangle** has three sides and three angles with a sum of 180°. The perimeter is the sum of all the side lengths.

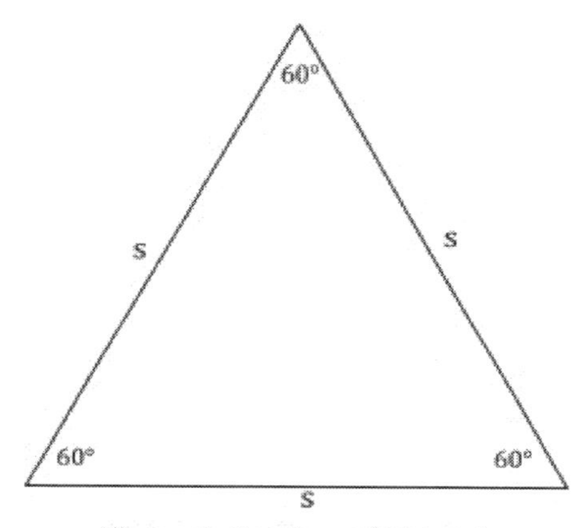

Figure 1: Equilateral Triangle

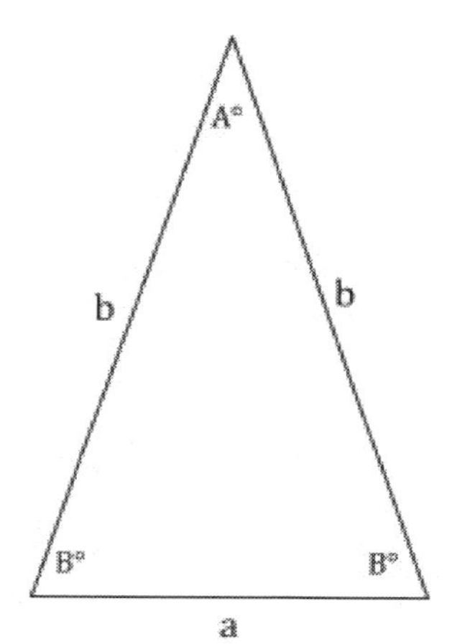

Figure 2: Isosceles Triangle

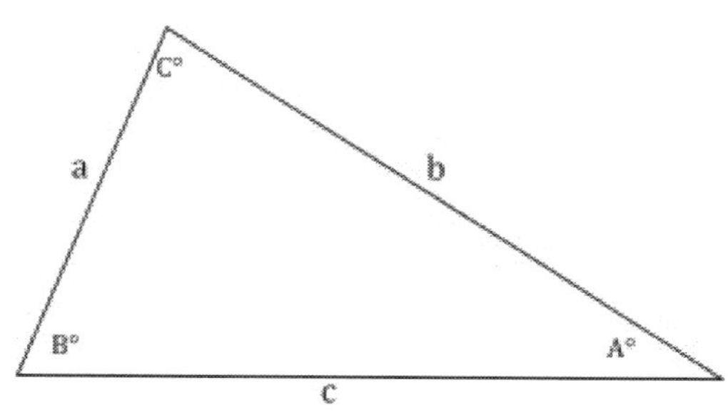

Figure 3: Scalene Triangle

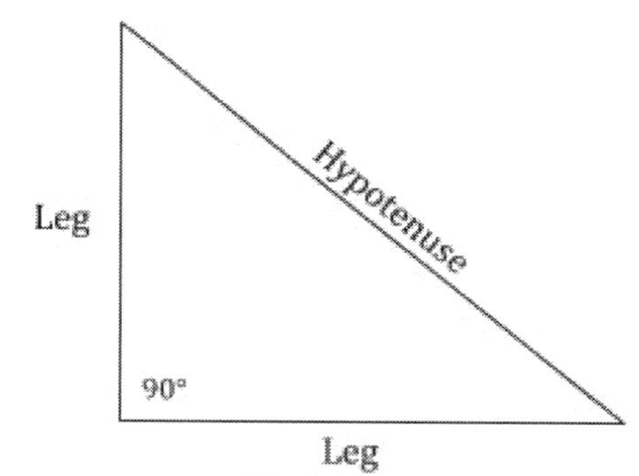

Figure 4: Right Triangle

Figure 1 is an **equilateral triangle.** An equilateral triangle is a triangle where all the sides are equal and all angles are 60 degrees. (Note: $60° + 60° + 60° = 180°$ or 180 degrees). The perimeter = $3s$, where s is the side length.

Figure 2 is an **isosceles triangle**. An isosceles triangle is a triangle where two sides and two angles are equal. The two equal sides are called the legs and the third side is called the base. The two equal angles are called the base angles and the third angle is called the vertex angle. Perimeter = $a + 2b$

Figure 3 is a **scalene triangle**. A scalene triangle is a triangle where all sides are different lengths and all angles are different measures. Perimeter = $a + b + c$

Figure 4 is a **right triangle**. A right triangle is a triangle with one right angle (90°). The side opposite of the right angle is called the hypotenuse and the other two sides are called the legs. Perimeter = $Leg_1 + Leg_2 +$ Hypotenuse

A **quadrilateral** has four sides and four angles with a sum of 360°.

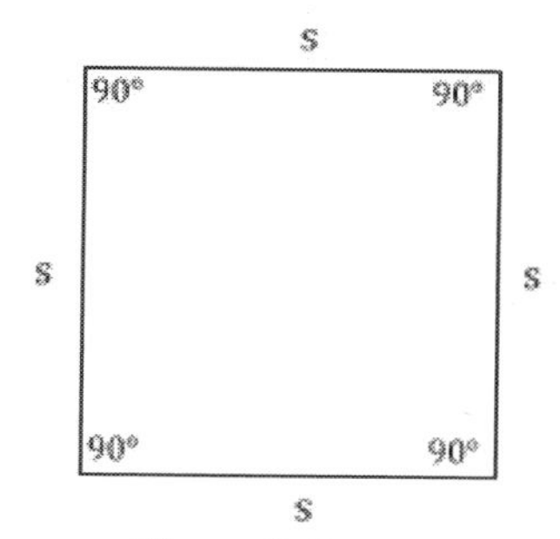

Figure 5: Square

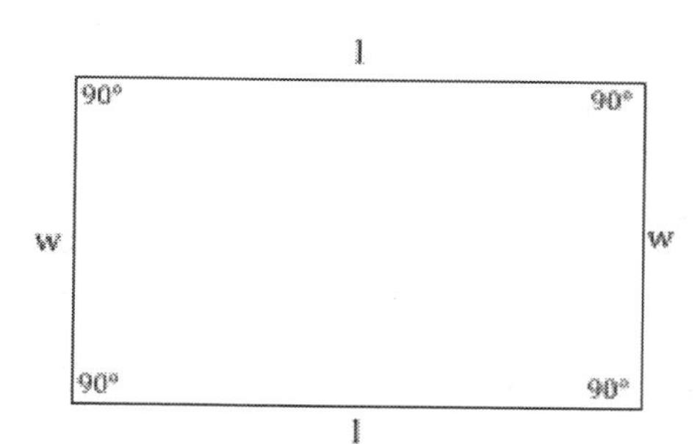

Figure 6: Rectangle

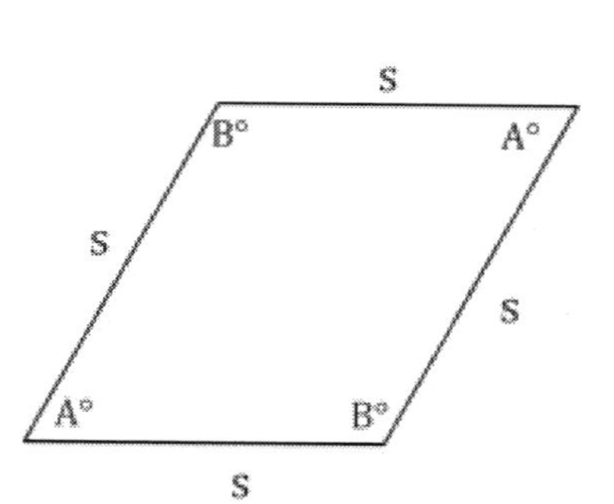

Figure 7: Rhombus

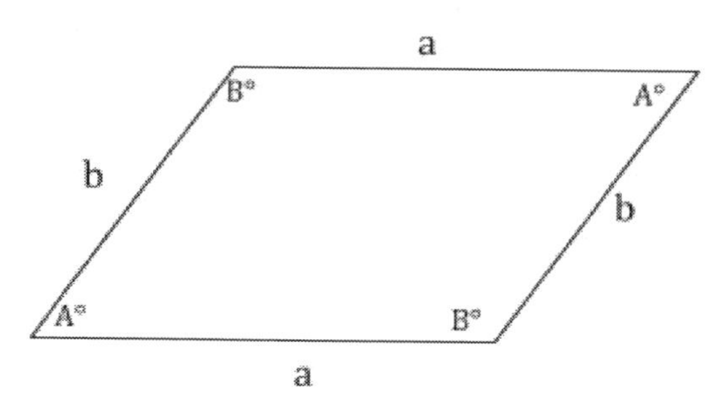

Figure 8: Parallelogram

Figure 5 is a **square** where all 4 sides are equal and all 4 angles are right angles (90 degrees). (Note: 90° + 90° + 90° + 90° = 360° or 360 degrees). The perimeter = $4s$, where *s* is the side length.

Figure 6 is a **rectangle** where opposite sides are equal and all 4 angles are equal or right angles (90 degrees). The perimeter = $2l + 2w$, where *l* is the length and *w* is the width.

Figure 7 is a **rhombus** where all 4 sides are equal and opposite angles are equal. The 2 different or adjacent angles have a sum of 180 degrees (A° + B° = 180°). The perimeter = $4s$, where *s* is the side length.

Figure 8 is a **parallelogram** where opposite sides are equal and opposite angles are equal. The 2 different angles have a sum of 180 degrees (A° + B° = 180°). The perimeter = $2a + 2b$, where *a* and *b* are side lengths.

Example 1: What is the third angle of a triangle in which two of its angles are 34° and 76°?

Solution: The 3 angles in a triangle have a sum of 180°.

Thus, subtract 34° and 76° from 180° to find the last angle.

180° – 34° – 76° = 70°

Answer: 70° or 70 degrees

Example 2: Find the perimeter of an equilateral triangle with side length 5.

Solution: By definition, an equilateral triangle has 3 sides that are equal in length.
Perimeter = 5 + 5 + 5 = 15

The formula, $p = 3s$, can also be used. $p = 3 * 5 = 15$

Answer: 15

Example 3: A square has a perimeter 48. What is the side length of the square?

Solution: The formula for the perimeter of a square is $p = 4s$.

Thus, $4s = 48$

Dividing both sides by 4 will yield the side length of the square:

$s = 12$

Answer: 12

Example 4: The length and width of a parallelogram is 18 and 12, respectively. What is its perimeter?

Solution: The formula for the perimeter of a parallelogram is p = $2a + 2b$.

$p = 2 * 18 + 2 * 12 = 36 + 24 = 60$

Answer: 60

Example 5: If one of a parallelogram's angles is 60°, find the other three angles.

Solution: Since opposite angles are equal, one of the three angles is also 60°.

The other angle is supplementary to 60° because different angles in a parallelogram add up to 180°.

Thus, 180° – 60° = 120°.

The other two angles must be 120° because opposite angles are equal.

Answer: 60°, 120°, 120°

Chapter 17

Review Problems

1. What are the three angles of an equilateral triangle?

2. Find the perimeter of an equilateral triangle with side length 7.

3. One of a right triangle's angles is $32°$. What are the measures of its other two angles?

4. Find the side length of an equilateral triangle with perimeter 93.

5. Find the perimeter of a scalene triangle with side lengths: 3, 5 and 7.

6. One of the base angles of an isosceles triangle is 64°. What are the other two angles?

7. Find the perimeter of a square with side length 13.

8. What are the measurements of all four angles of a square?

9. Find all of the side lengths of a square with perimeter 104.

10. A square has the same perimeter as an equilateral triangle. If the side length of the triangle is 24, what is the side length of the square?

11. Two equal angles are supplementary. Find one of the two angles.

12. A quadrilateral has a perimeter 80 and three of its sides have lengths 21, 25 and 18. What is the fourth side length?

13. Find all of the side lengths of a rhombus with perimeter 56.

14. Two equal angles are complementary. Find one of the two angles.

15. If one of a rhombus' angles is 23°, find the other three angles.

16. Find the length of a rectangle if the perimeter is 84 and the width is 13.

17. A parallelogram, a square and an equilateral triangle have the same perimeter. If the 2 different side lengths of the parallelogram are 84 and 120, what is the sum of the side length of the square and the side length of the equilateral triangle?

*18. The vertex angle of an isosceles triangle is 24°. If the base angle of the isosceles triangle is the same as one of the angles in a rhombus, find all of the angles of the rhombus.

*19. What is the supplement of the complement of the supplement of 148°?

20. A rhombus and a rectangle have the same perimeter. The side lengths of the rectangle are 13 and 85. What are the side lengths of the rhombus?

Geometric Figures II

This chapter will explore different geometric figures and how to calculate their respective areas.

Area: The amount of space inside a figure.

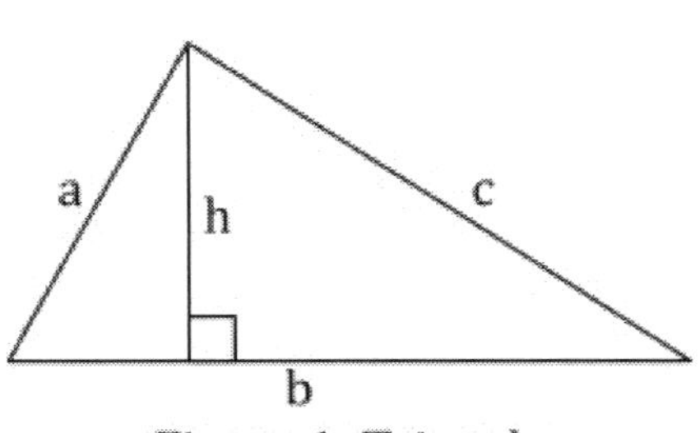

Figure 1: Triangle

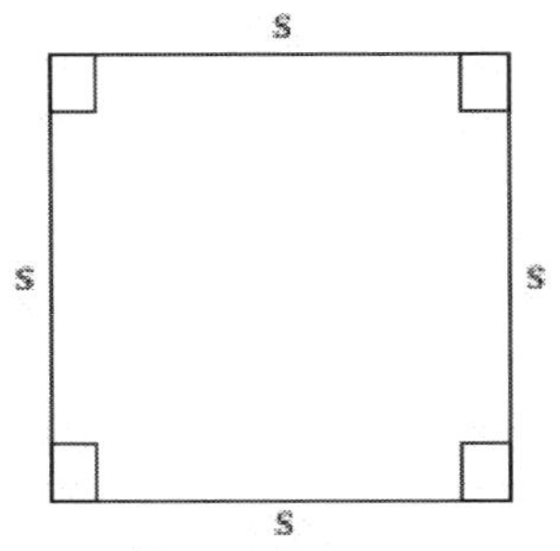

Figure 2: Square

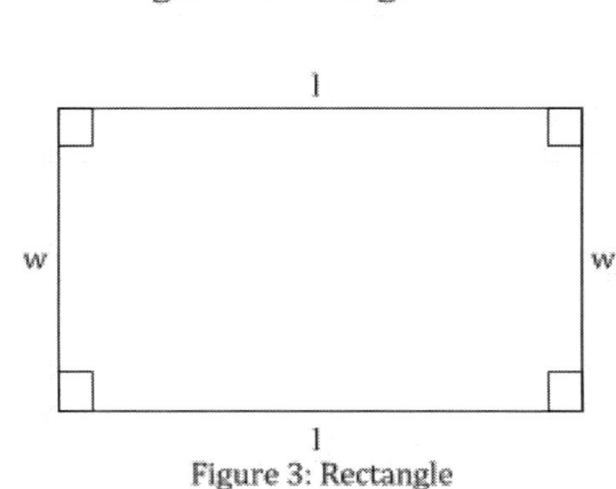

Figure 3: Rectangle

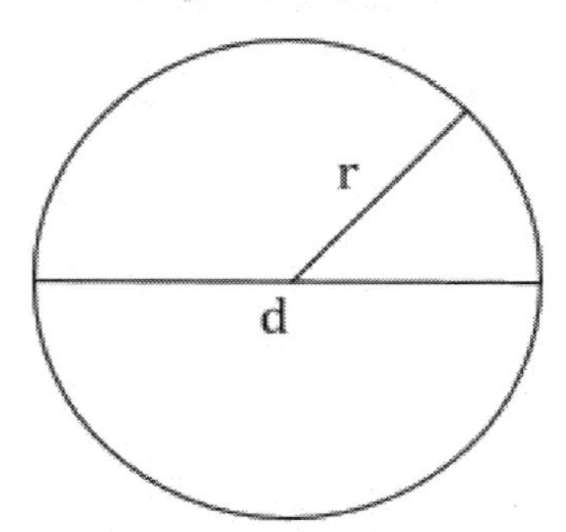

Figure 4: Circle

Figure 1 is a triangle.
The height, denoted by *h*, is the length from one vertex to the opposite side (in this case, side *b*) forming a right angle or 90 degree angle.

To calculate the area of a triangle, use the following formula: $A = \frac{1}{2}bh$
where *b* is the base and *h* is the height.

Thus, for a right triangle, the area is half the product of its two legs: $A = \frac{1}{2}l_1 l_2$

Figure 2 is a square.
To calculate the area of a square, use the following formula: $A = s^2 = s * s$
where *s* is the side length.

Figure 3 is a rectangle.
To calculate the area of a rectangle, use the following formula: $A = l * w$
where l is the length and w is the width.

Figure 4 is a circle. A circle is a shape that has the equal distance from its center. The distance from its center to any point on the circle is called the radius and is denoted by r. The diameter of a circle is equal to twice its radius. $d = 2r$ or Diameter = 2 × Radius.

Since a circle does not have straight sides, the total length around it is called the circumference.

The circumference of a circle is $C = \pi d$ or $C = 2\pi r$ where $d = 2r$.

$\pi = \frac{C}{d}$ $\qquad$ $\pi \approx 3.14$

π or pi is the ratio of the circumference to the diameter of a circle and is approximately equal to 3.14.

The area of a circle is $A = \pi r^2$, where r is the radius.

Example 1: Find the area of a triangle with base 12 and height 6. (Note: The height is perpendicular to the base)

Solution: Using $A = \frac{1}{2}bh$, $b = 12$ and $h = 6$,

$$A = \frac{1}{2} * 12 * 6 = 36$$

Answer: 36

Example 2: Find the area of a square with perimeter 12.

Solution: The perimeter of a square is $4s$. $P_{\text{square}} = 4s$
Thus, $4s = 12$. Dividing both sides by 4 yields: $s = 3$
Using the formula for the area of a square: $A = s^2$,
$A = 3^2 = 9$

Answer: 9

Example 3: Given a rectangle with area 15 where the side lengths are positive integers greater than 1, find its perimeter.

Solution: The prime factorization for 15 is 15 = 5 × 3, where 5 and 3 are both prime numbers. 3 must be the width and 5 must be the length because the other two factors are 1 and 15 but the side lengths must be greater than 1.

Thus, $l = 5$ and $w = 3$, Area = $A = lw = 5 * 3 = 15$

The perimeter formula for a rectangle is p = $2l + 2w$.

p = $2 * 5 + 2 * 3 = 16$

Answer: 16

Example 4: Find the circumference of a circle with radius 8. Express your answer in terms of π.

Solution: The formula for the circumference of a circle is: $C = 2\pi r$

Thus, $C = 2\pi * 8 = 16\pi$

Answer: 16π

Example 5: Find the area of a circle with diameter 14. Express your answer in terms of π.

Solution: The formula for the area of a circle is $A = \pi r^2$

Since the problem gives the diameter, the formula $d = 2r$ can be used to find the radius.

$d = 2r$ 14 = 2r

Divide both sides by 2: $r = 7$

Now, plug into the area formula: $A = \pi r^2 = \pi * 7^2 = 49\pi$

Answer: 49π

Review Problems

1. What is the area of a square with side length 10?

2. What is the length of a rectangle with area 80 and width 10?

3. What is the area of a triangle with base 16 and height 20? (Note: The height is perpendicular to the base)

4. What is the circumference of a circle with diameter 16? Express your answer in terms of π.

5. What is the area of a circle with diameter 22? Express your answer in terms of π.

6. A triangle and a square have the same area. If the base of the triangle is 12 and the side length of the square is 6, what is the height of the triangle? (Note: The height is perpendicular to the base)

7. A square and a rectangle have the same area. If the rectangle's length is 9 and its width is 4, what is the side length of the square?

*8. An isosceles right triangle has the same area as a rectangle with side lengths 10 and 5. What is the length of one of the legs of the right triangle?

9. Find the area of a rectangle with length 10 and perimeter 30.

10. What is the area of a right triangle with legs 11 and 60, and hypotenuse 61?

11. Find the circumference of a circle with area 81π. Express your answer in terms of π.

12. Given a rectangle with area 91 where the side lengths are positive integers greater than 1, find its perimeter.

13. A square's side length has the same length as the radius of a circle with circumference 24π. What is the area of the square?

14. What is the positive difference between the numerical value of circumference and the area of a circle with a diameter of 18? Express your answer in terms of π.

15. Find the area of the circle where its radius is equivalent to the width of rectangle with perimeter 26 and length 9. Express your answer in terms of π.

16. A rectangle's area is numerically equivalent to a square's perimeter. If the side lengths of the rectangle are 10 and 14, what is the side length of the square?

17. A right triangle has area 180. If one of the legs is 40, find the other leg.

18. A large square consists of 16 smaller squares (4 by 4) each with side length 5. Find the area of the large square.

*19. The first circle has diameter 16. The second circle has a circumference equivalent to the sum of the numerical value of the first circle's circumference and area. Find the area of the second circle.

*20. An isosceles triangle has the same area as a rectangle with length 12 and width 5. If the legs of the triangle are 13 and the height is 12, what is the perimeter of the triangle?

Chapter 19

Geometric Figures III

Polygons

Prefix = Meaning	**Shape**		**Number of Sides**
Tri = Three	Triangle	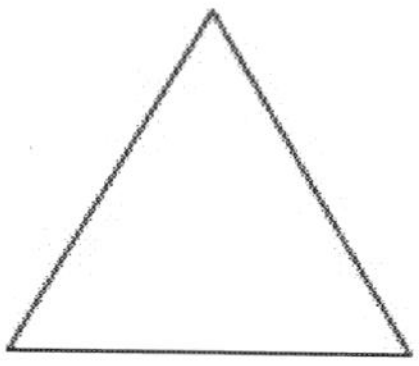	Three
Quad = Four	Quadrilateral	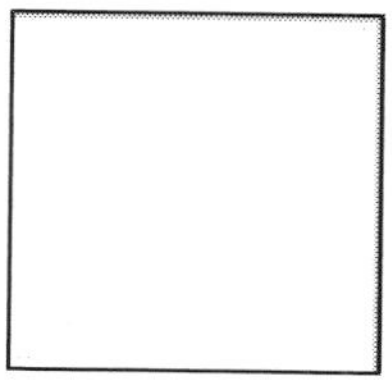	Four
Penta = Five	Pentagon	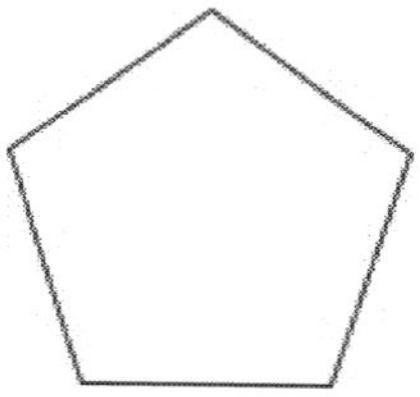	Five
Hexa = Six	Hexagon	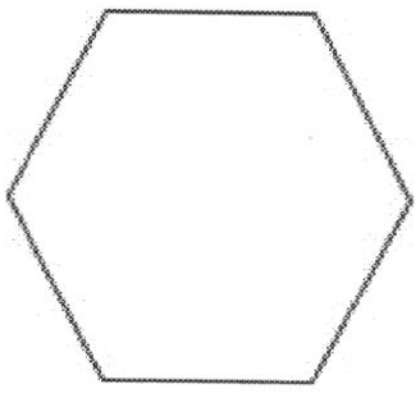	Six
Hepta = Seven	Heptagon	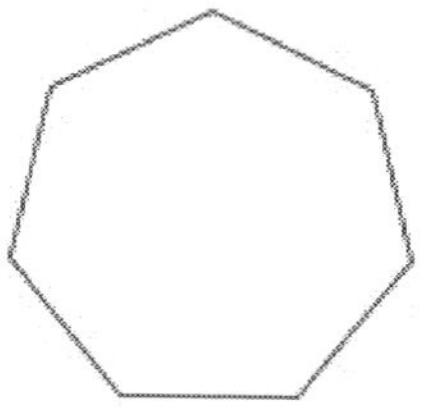	Seven

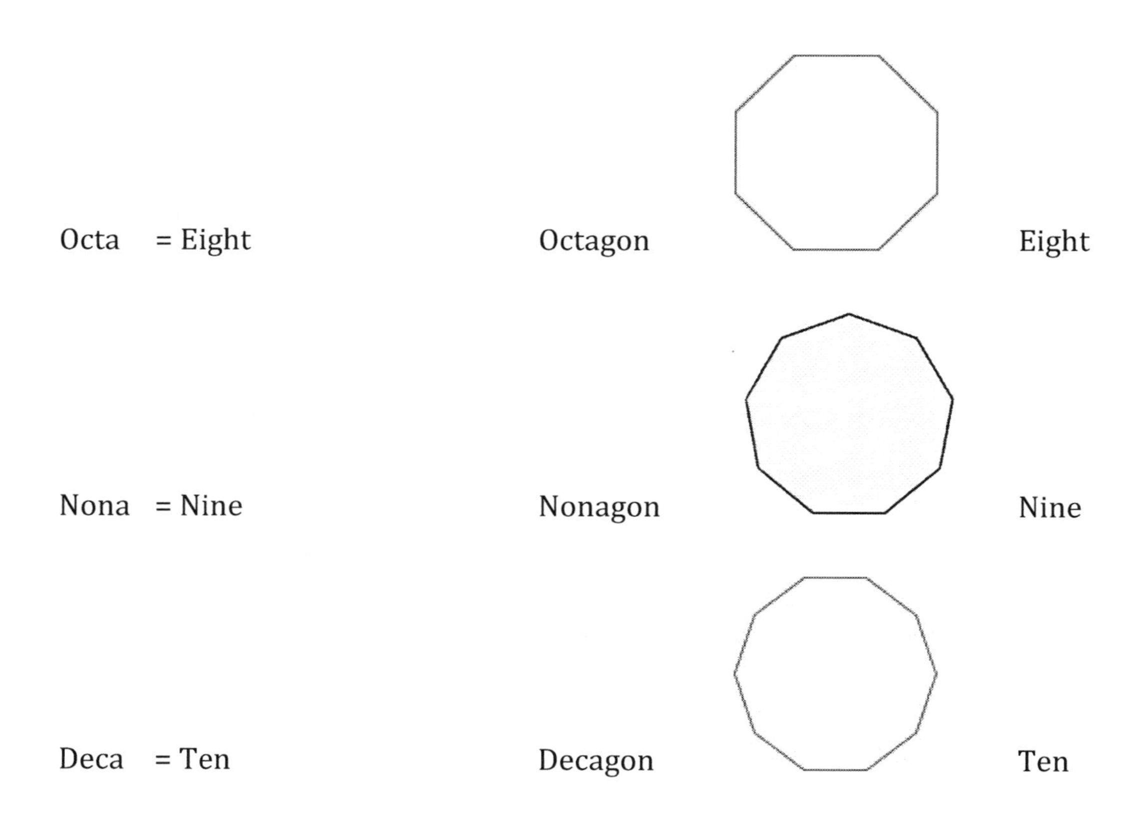

Figures with three or more sides are called **polygons**.
A **regular polygon** has angles and sides that are all equivalent.
An **equilateral polygon** has all of its sides equal and an **equiangular polygon** has all of its angles equal. Thus, a regular polygon is equilateral and equiangular.

For example, an equilateral triangle is a regular polygon because all of its angles are 60 degrees and all of its sides are of the same length.

Angles

The sum of all of the angles of a polygon is determined by the number of its sides. For a triangle, the sum of its angles is 180 degrees. For a quadrilateral, the sum of its angles is 360 degrees.

To calculate the sum of all the angles in an *n*-gon, where *n* is the number of its sides, use this formula: Sum = $(n-2)180$.

For a regular or equiangular polygon, to calculate one of its angles,

Use this formula: $\frac{(n-2)*180}{n} = 180 - \frac{360}{n}$

Perimeter

The perimeter of any polygon is the sum of all of its side lengths. For a regular *n*-gon with side length *s*, the perimeter formula is $p = n * s$.

Number of Diagonals

A diagonal is a line drawn from two vertices of a polygon that is not a side. A triangle has no diagonals and a quadrilateral has 2 diagonals.

Use this formula to calculate the number of diagonals for an *n*-gon: d = $\frac{n * (n - 3)}{2}$

Example 1: Find the perimeter of an equilateral hexagon with side length 4.

Solution: An equilateral hexagon has six equal sides. Thus, the perimeter is:
$p = n * s = 6 * 4 = 24$

Answer: 24

Example 2: Find one of the angles for an equiangular octagon.

Solution: An equiangular octagon has eight equal angles.

Find the sum of all of the angles of an octagon:
Sum = $(n - 2) * 180 = (8 - 2) * 180 = 6 * 180 = 1080$
Since each of the eight angles are equal, the sum can be divided by 8 to yield one of the angles.
1080 ÷ 8 = 135

Answer: 135 degrees

Example 3: Find the mean of all the angles in a triangle.

Solution: Sum = $(n - 2) * 180 = (3 - 2) * 180 = 180$
The sum of all the angles in a triangle is 180.

A triangle has 3 angles.

Thus, the average = $\frac{180}{3} = 60$

Note: This is equivalent to the one of the angles in a regular or equiangular triangle.

Answer: 60 degrees

Example 4: How many diagonals does a regular decagon have?

Solution: A regular decagon has 10 sides.

Number of diagonals: d = $\frac{n * (n-3)}{2} = \frac{10 * (10-3)}{2} = \frac{10 * 7}{2} = 35$

Answer: 35 diagonals

Example 5: An equilateral hexagon has the same side length as an equilateral nonagon. If the perimeter of the nonagon is 657, what is the perimeter of the hexagon?

Solution: An equilateral nonagon has 9 equal sides. Thus, one of the sides can be found by dividing the perimeter by 9.

$p = n * s \quad 657 = 9 * s$

$s = \frac{657}{9} = 73$

The hexagon and the nonagon both have side length 73. An equilateral hexagon has 6 equal sides. Thus, the perimeter is six times the side length.

$p = 6 * 73 = 438$

Answer: 438

Chapter 19

Review Problems

1. What is the perimeter of an equilateral quadrilateral with side length 5?

2. A triangle has perimeter 25 with side lengths 12 and 5. What is the third side length of the triangle?

3. Find one of the angles of an equiangular hexagon.

4. If three of the side lengths of a hexagon are 1 and the other three side lengths are 2, what is the perimeter of the hexagon?

5. Find one of the angles of a regular decagon.

6. Find the side length of an regular octagon with perimeter 24.

7. A hexagon has two different angle measures. Three of its angles are 100 degrees and the other three angles are X degrees. What is the value of X?

8. What is the sum of all the angles of a pentagon?

9. How many diagonals does a pentagon have?

10. Find the mean of all of the angles in a nonagon.

11. Find the positive difference between the side length of a regular pentagon with perimeter 35 and the side length of a regular heptagon with perimeter 21.

12. A decagon has side lengths: 1, 2, 3, 4, 5, 6, 7, 8, 9 and 10. If an equilateral pentagon has the same perimeter as the decagon, what is the side length of the pentagon?

13. How many diagonals does a heptagon have?

14. How many more diagonals does a 19-gon have than a 18-gon?

15. How many more diagonals does a 20-gon have than a 19-gon?

16. A regular quadrilateral, a regular pentagon, a regular hexagon and a regular heptagon have the same perimeter. If the side length of the hexagon is 105, what is the sum of the side lengths of the quadrilateral, pentagon, hexagon and heptagon?

17. Find the positive difference between one of the angles in a regular 20-gon and one of the angles in a regular 30-gon.

18. The numerical value of one of the angles in a regular octagon is equivalent to one of its side lengths. What is the perimeter of this octagon?

19. Find the sum of one of the angles in a regular 360-gon and one of the angles in a regular 180-gon.

20. A hexagon's angles form an arithmetic sequence. If the smallest angle is 95 degrees and the largest angle is 145 degrees, find the other four angles.

The Pythagorean Theorem

The pythagorean theorem is used to find the side lengths of right triangles. It states that given a triangle with legs a and b and hypotenuse c, the sum of the squares of the legs is equal to the square of the hypotenuse.

$$a^2 + b^2 = c^2$$

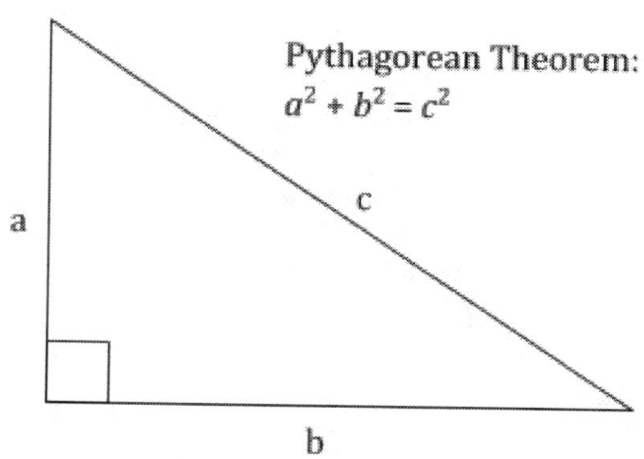

Figure 1: Right Triangle

Pythagorean Triples

Pythagorean triples are the three integers that make up the sides of a right triangle.

Pythagorean triples:

3–4–5	5–12–13	7–24–25
8–15–17	9–40–41	11-60-61

Difference of Two Squares

$$a^2 - b^2 = (a + b)(a - b)$$

Example 1: Find the hypotenuse of a right triangle with legs 3 and 4.

Solution: Let $a = 3$ and $b = 4$, then find c.

$c^2 = a^2 + b^2 = 3^2 + 4^2 = 9 + 16 = 25$

$c = \sqrt{25} = \sqrt{5^2} = 5$

Answer: 5

Example 2: Find the length of one of the diagonals of a rectangle with side lengths 5 and 12.

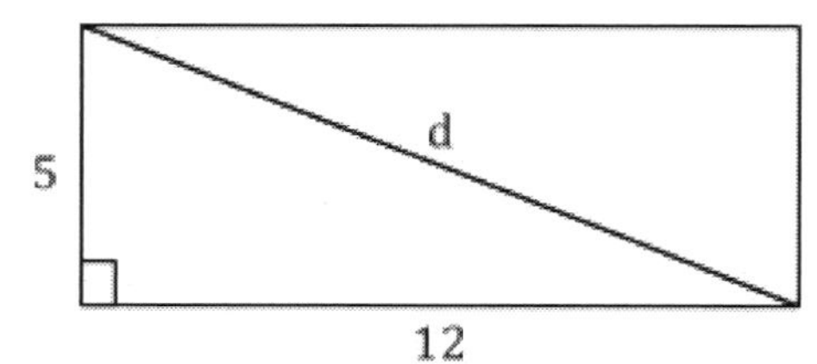

Solution:

Examining the figure above, one of the diagonals is the same as the hypotenuse of a right triangle with legs 5 and 12.

Thus, applying the pythagorean theorem:

$c^2 = a^2 + b^2 = 5^2 + 12^2 = 25 + 144 = 169$

$c = \sqrt{169} = \sqrt{13^2} = 13$

Answer: 13

Example 3: A square has a diagonal $4\sqrt{2}$. Find the area of the square.

Solution: The diagonal of the square, similar to the rectangle, will be the hypotenuse of the right triangle formed by two of its sides. In this case, the square has the same side lengths.

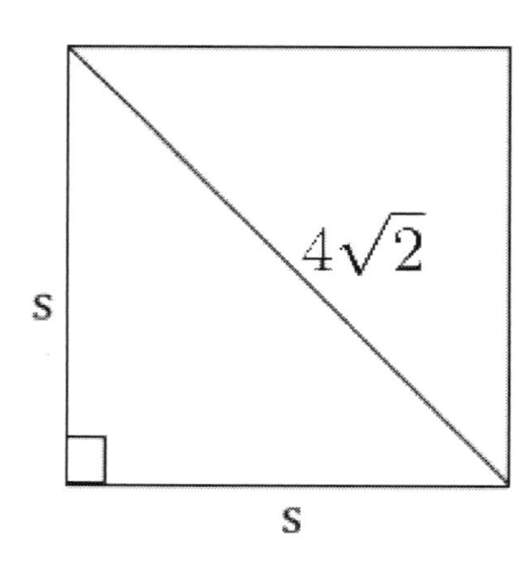

By the pythagorean theorem,

$(4\sqrt{2})^2 = s^2 + s^2$

$2s^2 = 16 * 2$

$s^2 = 16$

$s = \sqrt{16} = 4$

The area of a square is $A = s^2 = 4^2 = 4 \times 4 = 16$

Answer: 16

Example 4: A right triangle has leg 16 and hypotenuse 34. Find the other leg.

Solution: Using the pythagorean theorem,

$c^2 = a^2 + b^2$ becomes $34^2 = a^2 + 16^2$

$a^2 = 34^2 - 16^2$

Instead of squaring the two numbers and then subtracting, using the difference of two squares is quicker.

$a^2 = 34^2 - 16^2 = (34 + 16)(34 - 16) = 50 \times 18$

$a = \sqrt{50 * 18} = \sqrt{2^2 3^2 5^2} = 2 * 3 * 5 = 30$

Answer: 30

Example 5: Find the perimeter of the following quadrilateral:

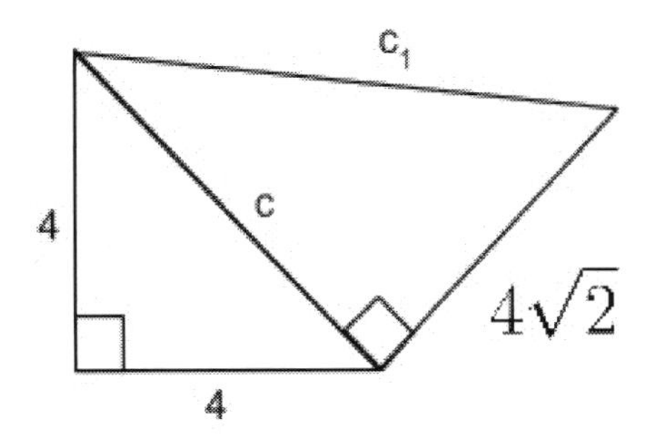

Figure not drawn to scale

Solution: To find the final side length of the quadrilateral, the pythagorean theorem must be applied twice.

The hypotenuse of the triangles with legs 4 and 4 must be found first.
$c^2 = a^2 + b^2 = 4^2 + 4^2 = 16 + 16 = 32$
$c = \sqrt{32} = 4\sqrt{2}$

The second right triangle has two legs $4\sqrt{2}$ and $4\sqrt{2}$.
Now, the hypotenuse of this triangle will yield the fourth side of the quadrilateral:
$c_1^2 = a_1^2 + b_1^2 = (4\sqrt{2})^2 + (4\sqrt{2})^2 = 32 + 32 = 64$
$c_1 = \sqrt{64} = 8$

All 4 sides of the quadrilateral have now been found. The perimeter can now be calculated:

$$p = 4 + 4 + 4\sqrt{2} + 8 = 16 + 4\sqrt{2}$$

Answer: $16 + 4\sqrt{2}$

Chapter 20

Review Problems

1. Find the hypotenuse of a right triangle with legs 6 and 8.

2. Find the hypotenuse of a right triangle with legs 300 and 400.

3. Find the hypotenuse of an isosceles right triangle with leg s. Express your answer in terms of s. Note: An isosceles right triangle has 2 equal legs.

4. Find the hypotenuse of an isosceles right triangle with legs 10 and 10.

5. Find the hypotenuse of a right triangle with legs 4 and 6.

6. Find the other leg of a right triangle with leg $\sqrt{15}$ and hypotenuse $\sqrt{31}$.

7. Find the perimeter of a right triangle with legs $4\sqrt{2}$ and $2\sqrt{2}$.

8. Find the area of a rectangle with a length $2\sqrt{55}$ and a diagonal 20.

9. Find the other leg of a right triangle with leg 84 and hypotenuse 85.

10. Find the area of a right triangle with leg 48 and hypotenuse 50.

11. Find the perimeter of a square with a diagonal $\sqrt{14}$.

12. Find the area of a square with a diagonal d. Express your answer in terms of d.

13. Find the area of a square with a diagonal $\sqrt{78}$.

14. Find the perimeter of the largest right triangle in the following figure. (Figure not drawn to scale)

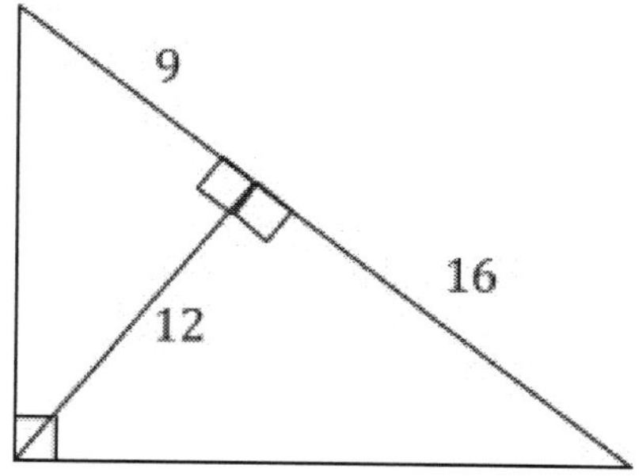

15. Find the perimeter of a right triangle with leg 11 and hypotenuse 61.

16. Find the area of the following quadrilateral (Figure not drawn to scale):

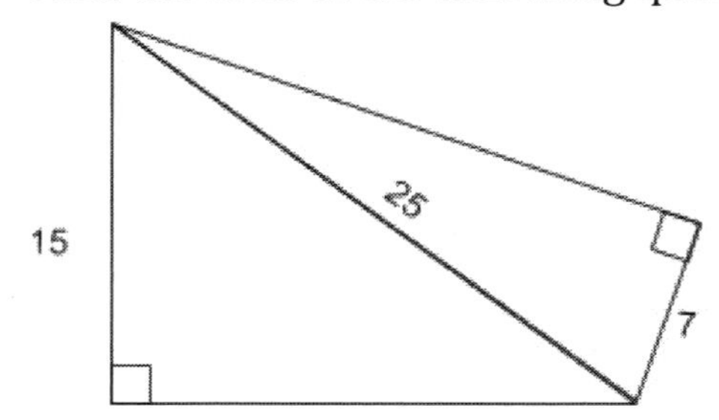

17. Find the perimeter of an isosceles right triangle with a leg s. Express your answer in terms of s. Hint: An isosceles right triangle has 2 equal legs.

18. Find the area of an isosceles right triangle with perimeter $28 + \sqrt{392}$. Hint: Use your answer from problem #17.

19. Find the sum of the numerical value of the perimeter and the area of a square with a diagonal $\sqrt{120}$.

20. Find the perimeter of the following pentagon. (Figure not drawn to scale)

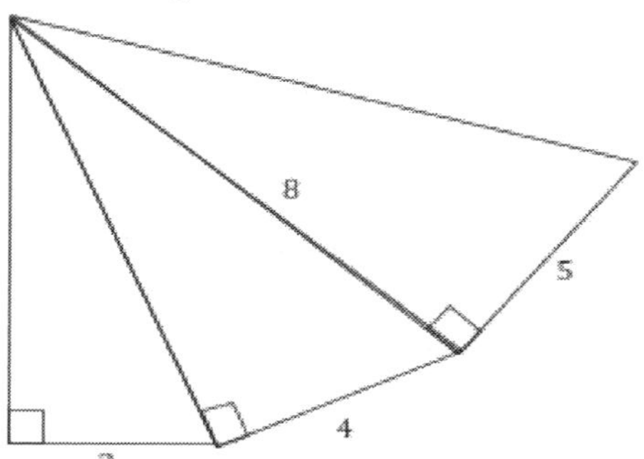

Review Problems

1. Solve: $1 - 2 + 3 - 4 + 5$

Solution: $1 - 2 + 3 - 4 + 5 = -1 + 3 - 4 + 5 = 2 - 4 + 5 = -2 + 5 = 3$

Answer: 3

2. Solve: $1 + 2 \times 2$

Solution: $1 + 2 \times 2 = 1 + 4 = 5$

Answer: 5

3. Solve: $1 - (2 + 3) - (4 + 5)$

Solution: $1 - (2 + 3) - (4 + 5) = 1 - 5 - 9 = -4 - 9 = -13$

Answer: −13

4. Solve: $(1 + 2) \times 2$

Solution: $(1 + 2) \times 2 = 3 \times 2 = 6$

Answer: 6

5. Solve: $12 \div 2 \times 3$

Solution: $12 \div 2 \times 3 = 6 \times 3 = 18$

Answer: 18

6. Solve: $0 + 2 - 4 \times 6 \div 8$

Solution: $0 + 2 - 4 \times 6 \div 8 = 0 + 2 - 24 \div 8 = 0 + 2 - 3 = 2 - 3 = -1$

Answer: −1

7. Solve: $1 \times 2 + 3 \times 4 + 5 \times 6 + 7 \times 8$

Solution:
$1 \times 2 + 3 \times 4 + 5 \times 6 + 7 \times 8 =$
$2 + 3 \times 4 + 5 \times 6 + 7 \times 8 =$
$2 + 12 + 5 \times 6 + 7 \times 8 =$
$2 + 12 + 30 + 56 =$
$14 + 30 + 56 =$
$44 + 56 = 100$

Answer: 100

8. Solve: $1 \times (2 + 3) \times 4 + 5 \times 6 + 7 \times 8$

Solution:
$1 \times (2 + 3) \times 4 + 5 \times 6 + 7 \times 8 =$
$1 \times 5 \times 4 + 5 \times 6 + 7 \times 8 =$
$5 \times 4 + 5 \times 6 + 7 \times 8 =$
$20 + 5 \times 6 + 7 \times 8 =$
$20 + 30 + 7 \times 8 =$
$20 + 30 + 56 = 50 + 56 = 106$

Answer: 106

9. How many integers are between -10 and 10?

Solution: Using the formula, $(b - a) - 1$:
$(10 - (-10)) - 1 = (10 + 10) - 1 = 20 - 1 = 19$

Answer: 19 integers

10. How many natural numbers are between $-12{,}394$ and 932, inclusive?

Solution: The first natural number is 1 and the greatest natural number in this set is 932.
Using the formula, $(b - a) + 1$:
$(932 - 1) + 1 = 931 + 1 = 932$

Answer: 932 natural numbers

11. Solve: $32 - (16 - (8 - (4 - (2 - 1))))$
Hint: Start with the parentheses that is nested the most.

Solution: $32 - (16 - (8 - (4 - (2 - 1)))) =$
$32 - (16 - (8 - (4 - 1))) =$
$32 - (16 - (8 - 3)) =$
$32 - (16 - 5) =$
$32 - 11 = 21$

Answer: 21

12. Solve: $2^{((2 + 3) \times 4 - 18)}$

Solution: $2^{((2 + 3) \times 4 - 18)} = 2^{(5 \times 4 - 18)} = 2^{(20 - 18)} = 2^{(2)} = 4$

Answer: 4

13. Solve: $6(2 + 4)$
    Hint: $6(2 + 4) = 6 \times (2 + 4)$.

Solution: $6(2 + 4) = 6 \times (2 + 4) = 6 \times 6 = 36$

Answer: 36

14. Solve: $20(50 + 4) - 1000$

Solution: $20(50 + 4) - 1000 = 20 \times (50 + 4) - 1000 = 20 \times 54 - 1000 = 1080 - 1000 = 80$

Answer: 80

15. Solve: $5(2 + 4 + 6 + 8 + 10)$

Solution: $5(2 + 4 + 6 + 8 + 10) = 5 \times (2 + 4 + 6 + 8 + 10) = 5 \times (30) = 150$

Answer: 150

16. Solve: $32 \div 16 \times 6 \div 3 \times 10 \div 5 + 18 \div 9 \times 12 \div 6 \times 8 \div 4$

Solution: $32 \div 16 \times 6 \div 3 \times 10 \div 5 + 18 \div 9 \times 12 \div 6 \times 8 \div 4 =$
$2 \times 6 \div 3 \times 10 \div 5 + 2 \times 12 \div 6 \times 8 \div 4 =$
$12 \div 3 \times 10 \div 5 + 24 \div 6 \times 8 \div 4 =$
$4 \times 10 \div 5 + 4 \times 8 \div 4 =$

$40 \div 5 + 32 \div 4 =$
$8 + 8 = 16$

Answer: 16

17. How many integers are between –34 and 89, inclusive?

Solution: Using the formula, $(b - a) + 1$:
$(89 - (-34)) + 1 = (89 + 34) + 1 = 123 + 1 = 124$

Answer: 124 integers

18. Jake has to read his book from page 186 to page 563, inclusive. If he wants to read the same number of pages for 14 days, how many pages should he read each day?

Solution: Using the formula, $(b - a) + 1$:
$(563 - 186) + 1 = 377 + 1 = 378$
$378 \div 14 = 27$

Answer: 27 pages

19. Solve: $16 \div (2 + 2)^2 + (19 - 2 \div 2) \div 6$

Solution:
$16 \div (2 + 2)^2 + (19 - 2 \div 2) \div 6 =$
$16 \div 4^2 + (19 - 1) \div 6 =$
$16 \div 16 + 18 \div 6 = 1 + 3 = 4$

Answer: 4

20. Solve: $720 \div 6 \div 5 \div 4 \div 3 \div 2 \div 1$

Solution:
$720 \div 6 \div 5 \div 4 \div 3 \div 2 \div 1 =$
$120 \div 5 \div 4 \div 3 \div 2 \div 1 =$
$24 \div 4 \div 3 \div 2 \div 1 =$
$6 \div 3 \div 2 \div 1 =$
$2 \div 2 \div 1 =$
$1 \div 1 = 1$

Answer: 1

Review Problems

1. Which of these numbers are divisible by 5? {365, 549, 766, 450, 324}

Solution: The divisibility rule for 5 is if the number ends in 0 or 5.

Answer: 365 and 450

2. Is 108 divisible by 6?

Solution: The divisibility rule for 6 is if the number is divisible by 2 and 3.
108 is even, thus it is divisible by 2.
The sum of the digits is $1 + 0 + 8 = 9$, which is divisible by 3.
Thus, 108 is divisible by 6.

Answer: Yes

3. How many whole numbers between 1 and 50 are divisible by 3?

Solution: $50 \div 3 = 16$ R 2.
This means that the largest multiple of 3 less than 50 is $3 \times 16 = 48$.
There are 16 multiples of 3 between 1 and 50 ($3 \times 1, 3 \times 2, \ldots, 3 \times 16$).

Answer: 16 (numbers)

4. Is 8,715 divisible by 15? Hint: $15 = 5 \times 3$

Solution: Since $15 = 5 \times 3$, if 8,715 is divisible by 5 and 3, it is divisible by 15.
8,715 ends in 5 which means it is divisible by 5.
The sum of the digits is $8 + 7 + 1 + 5 = 21$, which is divisible by 3.
Therefore, 8,715 is divisible by 15.

Answer: Yes

5. What is the largest two-digit whole number that is divisible by 6?

Solution: The largest two-digit whole number is 99.
$99 \div 6 = 16$ R 3.
Thus, $6 \times 16 = 96$ is the largest two-digit whole number that is divisible by 6.

Answer: 96

6. What are the remainders when 123,456 is divided by 4, 5, 8, 9, or 10?

Solution:
4: Last two digits: $56 \div 4 = 14$ R **0**
5: Last digit: $6 \div 5 = 1$ R **1**
8: Last three digits: $456 \div 8 = 57$ R **0**
9: Sum of digits: $1 + 2 + 3 + 4 + 5 + 6 = 21$; $21 \div 9 = 2$ R **3**
10: Last digit: $6 \div 10 = 0$ R **6**

Answer: 0, 1, 0, 3, 6

7. How many numbers between 1 and 100 are divisible by 7?

Solution: $100 \div 7 = 14$ R 2.
This means that the largest multiple of 7 less than 100 is $7 \times 14 = 98$.
There are 14 multiples of 7 between 1 and 100 ($7 \times 1, 7 \times 2, \ldots, 7 \times 14$).

Answer: 14 (numbers)

8. If the three-digit number 4A6 is divisible by 9, what is the value of A?

Solution: Sum of digits: $4 + A + 6 = 10 + A$. This value must be equal to a multiple of 9.
$10 + A = 9$; $A = -1$ $\qquad$ $10 + A = 18$; $A = 8$ $\qquad$ $10 + A = 27$; $A = 17$
Since A is a digit (0–9), the only value that works is 8. A must be 8.

Answer: 8

9. Which of these numbers is divisible by 8? {25,412, 34,612, 78,512, 56,388}

Solution: Only the last three digits need to be check.
$412 \div 8 = 51$ R 4 $\qquad$ $612 \div 8 = 76$ R 4 $\qquad$ $512 \div 8 = 64$ R 0 $\qquad$ $388 \div 8 = 48$ R 4
512 is the only number divisible by 8. Thus, 78,512 is divisible by 8.

Answer: 78,512

*10. The six-digit number 8A4,B57 is divisible by 11. What is the sum of A and B?

Solution: The difference between the sums of alternating digits must be calculated.

Even-numbered digits: $8 + 4 + 5 = 17$
Odd-numbered digits: $A + B + 7$

$(A + B + 7) - 17 = A + B - 10$. This value must be divisible by 11 or equal to 0.
$A + B - 10 = -11$; $A + B = -1$ $\quad$ $A + B - 10 = 0$; $A + B = 10$
$A + B - 10 = 11$; $A + B = 21$
Since A and B are digits (0–9), the sum must range from 0 to $9 + 9 = 18$. 10 is the only one that fits in that range. Thus, $A + B = 10$.

Answer: 10

11. Jerry is thinking of a number between 1 and 100. The number is a multiple of 11 and is also a multiple of 5. What number is he thinking of?

Solution: If a number is divisible by 5 and 11, the number must be divisible by $5 \times 11 = 55$. The only multiple of 55 between 1 and 100 is 55.

Answer: 55

*12. If 5,A3B is a multiple of 9, what is the sum of all possible values of A + B?

Solution: Sum of digits: $5 + A + 3 + B = 8 + A + B$. This value must be equal to a multiple of 9.
$8 + A + B = 9$; $A + B = 1$
$8 + A + B = 18$; $A + B = 10$
$8 + A + B = 27$; $A + B = 19$
Since A and B are digits (0–9), the sum can be from 0 to $9 + 9 = 18$, inclusive. Thus, only 1 and 10 work. $1 + 10 = 11$.

Answer: 11

13. What is the remainder when 62,579,304 is divided by 9?

Solution: Sum of digits: $6 + 2 + 5 + 7 + 9 + 3 + 0 + 4 = 36$.
$36 \div 9 = 4$ R **0**.

Answer: 0

14. How many multiples of 3 are between 11 and 302?

Solution: $11 \div 3 = 3$ R 2 and $302 \div 3 = 100$ R 2.
Thus, the smallest multiple of 3 greater than 11 is $3 \times (3 + 1) = 3 \times 4$ because 3×3 is less than 11. The greatest multiple of 3 less than 302 is 3×100.
The numbers are $(3 \times 4, 3 \times 5, \ldots, 3 \times 100)$
From chapter 1 of counting numbers inclusive, there are $(100 - 4) + 1 = 97$ multiples.

Answer: 97 (multiples)

15. The five-digit number A5,2A1 is divisible by 3. If the five-digit number is to be as large as possible, what is the value of A?

Solution: Sum of digits: $A + 5 + 2 + A + 1 = 2A + 8$. Since, the largest digit that A could be is 9, check to see if the sum is divisible by 3 and then work down.

Let $A = 9$: $2 \times 9 + 8 = 26$; 26 is not divisible by 3
Let $A = 8$: $2 \times 8 + 8 = 24$; 24 is divisible by 3: making 85,281 the largest possible number.

Answer: 8

16. What is the smallest three-digit number that is a multiple of 5?

Solution: The smallest three digit number is 100. Since 100 ends in 0, it is a multiple of 5.

Answer: 100

17. What is largest two-digit number that is divisible by 7 and 6?

Solution: If a number is divisible by 7 and 6, the number must be divisible by $7 \times 6 = 42$. The only multiples of 42 between 1 and 100 is 42 and 84. Thus 84 is the largest two-digit number that is a multiple of 7 and 6.

Answer: 84

18. Find the remainder when 7,649,432,789 is divided by 9.

Solution: Sum of digits: 7 + 6 + 4 + 9 + 4 + 3 + 2 + 7 + 8 + 9 = 59.
59 ÷ 9 = 6 R **5**

Answer: 5

19. Marina is trying to remember her favorite number. She remembers that the number is more than 40 and less than 70 and is divisible by 8 and 7. What is her favorite number?

Solution: If a number is divisible by 8 and 7, the number must be divisible by 8 × 7 = 56. The only multiple of 56 between 40 and 70 is 56. Thus, her favorite number is 56.

Answer: 56

*20. A number is randomly chosen from 1 to 100, inclusive. What is the probability that the number is divisible by 5 and 3? Express your answer as a percent.

Solution: If a number is divisible by 5 and 3, the number must be divisible by 5 × 3 = 15. 100 ÷ 15 = 6 R 10. This means there are 6 multiples of 15 from 1 to 100, inclusive.

$$\frac{6}{100} * 100 = 6 \text{ percent}$$

Answer: 6%

Review Problems

1. What is the sum of the first 4 prime numbers?

Solution: The first four prime numbers are 2, 3, 5 and 7. $2 + 3 + 5 + 7 = 17$.

Answer: 17

2. Is 84 a prime number? If not, find its prime factors.

Solution: 84 is **not** a prime number because it is even or divisible by 2.
$84 = 2 \times 42 = 2 \times 2 \times 21 = 2 \times 2 \times 3 \times 7$

Answer: No. 2, 3 and 7.

3. What is the largest prime number less than 100?

Solution: Start with the largest odd number less than 100 because all positive even numbers (except 2) are composite numbers. The only prime numbers that need to be checked are 3, 5 and 7 since the numbers are odd.

$99 \div 3 = 33$ R 0. 99 is composite.
$97 \div 3 = 32$ R 1. $97 \div 5 = 19$ R 2. $97 \div 7 = 13$ R 6.
97 is a prime number because it is not divisible by 2, 3, 5 or 7.

Answer: 97

4. How many prime numbers are divisible by 2?

Solution: 2 is the only prime number that is divisible by 2 or is even. The rest of the even positive numbers are composite numbers.

Answer: 1 (number)

5. What are the prime factors of 40?

Solution: $40 = 2 \times 20 = 2 \times 2 \times 10 = 2 \times 2 \times 2 \times 5$

Answer: 2 and 5

6. Is 119 a prime number? If not, find its prime factors. (Hint: 121 = 11 × 11.)

Solution: Since 121 is greater than 119, the only prime numbers that need to be checked are 2, 3, 5 and 7.
119 ÷ 2 = 59 R 1 119 ÷ 3 = 39 R 2 119 ÷ 5 = 23 R 4
119 ÷ 7 = 17 R 0. 119 is a composite number
119 = 7 × 17

Answer: No. 7 and 17.

7. Is 163 a prime number? If not, find its prime factors. (Hint: 169 = 13 × 13.)

Solution: Since 169 is greater than 163, the only prime numbers that need to be checked are 2, 3, 5, 7 and 11.
163 ÷ 2 = 81 R 1 163 ÷ 3 = 54 R 1 163 ÷ 5 = 32 R 3
163 ÷ 7 = 23 R 2 163 ÷ 11 = 14 R 9
Thus, 163 is prime.

Answer: Yes, 163 is a prime number

8. What prime numbers end in 5?

Solution: According to the divisibility rule for 5, if a number ends in 5, it is divisible by 5. 5 is the only prime number that ends in 5.

Answer: 5

9. How many prime numbers are between 1 and 50?

Solution: Using the Sieve of Eratosthenes, the only prime numbers that need to be checked are 2, 3, 5, and 7.
The prime numbers found are: 2, 3, 5, 7, 11, 13, 17, 19, 23, 29, 31, 37, 41, 43, 47 which gives a total of 15 prime numbers.

Answer: 15 (numbers)

10. Find the sum of the prime numbers between 20 and 50.

Solution: Using the prime numbers found in problem #9,
23 + 29 + 31 + 37 + 41 + 43 + 47 = 251

Answer: 251

11. What prime factors do 60 and 36 share?

Solution:
60 = 2 × 30 = 2 × 2 × 15 = 2 × 2 × 3 × 5
36 = 2 × 18 = 2 × 2 × 9 = 2 × 2 × 3 × 3
These numbers share 2 and 3 as prime factors.

Answer: 2 and 3

12. Find the sum of all prime numbers less than 100 that end in 7.

Solution: There are 10 numbers that end in 7: 7, 17, 27, 37, 47, 57, 67, 77, 87, 97
Now, check to see which numbers are divisible by 2, 3, 5 or 7.
2: None because they are all odd numbers
3: 27, 57, and 87 (sum of digits are divisible by 3)
5: None because they all end in 7
7: 7 and 77 except that 7 is a prime number
The prime numbers are thus: 7, 17, 37, 47, 67, and 97.
7 + 17 + 37 + 47 + 67 + 97 = 272.

Answer: 272

*13. What is the largest prime number less than 190 that ends in 7?

Solution: Start with 187 then work down. The only prime numbers that need to be checked are 3, 7, 11 and 13 because the number ends in 7 so it is not divisible by 2 or 5.
187 = 11 × 17
177 = 3 × 59
167 is not divisible by 3, 7, 11 or 13. Thus, it is a prime number.

Answer: 167

14. What prime factors do 135 and 90 share?

Solution: $135 = 5 \times 27 = 3 \times 5 \times 9 = 3 \times 3 \times 3 \times 5$
$90 = 2 \times 45 = 2 \times 3 \times 15 = 2 \times 3 \times 3 \times 5$
They share 3 and 5 as prime factors.

Answer: 3 and 5

15. Find the sum of all prime numbers less than 100 that end in 1.

Solution: There are 10 numbers that end in 1: 1, 11, 21, 31, 41, 51, 61, 71, 81, 91
1 is not a prime number. Now, check to see which numbers are divisible by 2, 3, 5 or 7.
2: None because they are all odd numbers
3: 21, 51, and 81 (sum of digits are divisible by 3)
5: None because they all end in 1
7: 21 and 91
The prime numbers are: 11, 31, 41, 61, 71
$11 + 31 + 41 + 61 + 71 = 215$

Answer: 215

*16. Is 1001 a prime number? If not, find its prime factors.

Solution: Check the first couple prime numbers: 2, 3, 5, 7
1001 ends in 1 so it is not divisible by 2 or 5. $1 + 0 + 0 + 1 = 2$ which is not divisible by 3.
$1001 \div 7 = 143$ R 0; $1001 = 7 \times 143$
Now check to see if 143 is a prime number. It ends in 3 so it is not divisible by 2 or 5. $1 + 4 + 3 = 8$ which is not divisible by 3.
$143 \div 7 = 20$ R 3; $143 \div 11 = 13$ R 0: $143 = 11 \times 13$
Thus, $1001 = 7 \times 143 = 7 \times 11 \times 13$

Answer: No. 7, 11 and 13.

*17. Find the prime factors of 183,183. Hint: $183{,}183 = 183 \times 1001$

Solution: $183{,}183 = 183 \times 1001 = 183 \times 7 \times 11 \times 13$
183 is divisible by 3. $183 \div 3 = 61$ R 0. 61 is a prime number.
$183 = 3 \times 61$
$183{,}183 = 183 \times 1001 = 183 \times 7 \times 11 \times 13 = 3 \times 61 \times 7 \times 11 \times 13$

Answer: 3, 7, 11, 13 and 61

18. What is the product of the first three prime numbers?

Solution: The first three prime numbers are 2, 3 and 5. 2 × 3 × 5 = 30

Answer: 30

*19. What is the greatest prime number less than 200? Hint: Work your way down from the greatest **odd** number and check if it's a prime number.

Solution: Starting with 199, then work down if needed. The only prime numbers that need to be tested are 2, 3, 5, 7, 11, and 13. (17 × 17 = 289 > 200)
199 ends in 9 so it is not divisible by 2 or 5.
1 + 9 + 9 = 19 which is not divisible by 3.
199 ÷ 7 = 28 R 3   199 ÷ 11 = 18 R 1   199 ÷ 13 = 15 R 4
Thus, 199 is a prime number.

Answer: 199

*20. Which of the numbers between 100 and 110 are prime numbers?

Solution: The only numbers that need to be checked are 101, 103, 107 and 109 because they are the only odd numbers and 105 is divisible by 5. The only prime numbers that need to be tested are 3 and 7 because they are not even or end in 0 or 5 so they are not divisible by 2 or 5.

101 ÷ 3 = 33 R 2;   101 ÷ 7 = 14 R 3
103 ÷ 3 = 34 R 1;   103 ÷ 7 = 14 R 5
107 ÷ 3 = 35 R 2;   107 ÷ 7 = 15 R 2
109 ÷ 3 = 36 R 1;   109 ÷ 7 = 15 R 4

Thus, all four of them are prime numbers.

Answer: 101, 103, 107 and 109

Review Problems

1. Find the prime factorization for 17.

Solution: 17 is a prime number. Thus, 17^1 is the prime factorization.

Answer: $\mathbf{17^1}$

2. Find the prime factorization for 91.

Solution: $91 = 7 \times 13 = 7^1 13^1$

Answer: $\mathbf{7^1 13^1}$

3. Find the number of factors for 6.

Solution: $6 = 2 \times 3 = 2^1 3^1$. The number of factors is $(1 + 1) \times (1 + 1) = 2 \times 2 = 4$.

Answer: 4 (factors)

4. Find the number of the factors for 65.

Solution: $65 = 5 \times 13 = 5^1 13^1$. The number of factors is $(1 + 1) \times (1 + 1) = 2 \times 2 = 4$.

Answer: 4 (factors)

5. Find the sum of all the factors for 30.

Solution: $30 = 2 \times 3 \times 5 = 2^1 3^1 5^1$. The sum of the factors is
$(2^0 + 2^1) \times (3^0 + 3^1) \times (5^0 + 5^1) = (1 + 2) \times (1 + 3) \times (1 + 5) = 3 \times 4 \times 6 = 72$

Answer: 72

6. Find the sum of all the factors for 77.

Solution: $77 = 7 \times 11 = 7^1 11^1$. The sum of the factors is
$(7^0 + 7^1) \times (11^0 + 11^1) = (1 + 7) \times (1 + 11) = 8 \times 12 = 96$

Answer: 96

7. Find the number of the factors for 84.

Solution: $84 = 2 \times 2 \times 3 \times 7 = 2^2 3^1 7^1$. The number of factors is $(2 + 1) \times (1 + 1) \times (1 + 1) = 3 \times 2 \times 2 = 12$

Answer: 12 (factors)

8. Find the prime factorization for 60.

Solution: $60 = 2 \times 2 \times 3 \times 5 = 2^2 3^1 5^1$.

Answer: $\mathbf{2^2 3^1 5^1}$

9. How many more factors does 60 have than 30?

Solution:
$60 = 2 \times 2 \times 3 \times 5 = 2^2 3^1 5^1$. The number of factors: $(2 + 1) \times (1 + 1) \times (1 + 1) = 12$
$30 = 2 \times 3 \times 5 = 2^1 3^1 5^1$. The number of factors: $(1 + 1) \times (1 + 1) \times (1 + 1) = 8$

$12 - 8 = 4$

Answer: 4 (factors)

10. Find the prime factorization for 8000.

Solution: $8000 = 2 \times 2 \times 2 \times 2 \times 2 \times 2 \times 5 \times 5 \times 5 = 2^6 5^3$.

Answer: $\mathbf{2^6 5^3}$

11. Find the prime factorization and the number of factors for 480.

Solution: $480 = 2 \times 2 \times 2 \times 2 \times 2 \times 3 \times 5 = 2^5 3^1 5^1$.
The number of factors is $(5 + 1) \times (1 + 1) \times (1 + 1) = 6 \times 2 \times 2 = 24$

Answer: $\mathbf{2^5 3^1 5^1}$ **and 24 (factors)**

12. Find the sum of all the factors for 8, 16, 32 and 64. List each sum separately.

Solution:
$8 = 2 \times 2 \times 2 = 2^3$. Sum of factors: $(2^0 + 2^1 + 2^2 + 2^3) = 1 + 2 + 4 + 8 = 15$.

$16 = 2 \times 2 \times 2 \times 2 = 2^4$. Sum of factors: $(2^0 + 2^1 + 2^2 + 2^3 + 2^4) = (2^0 + 2^1 + 2^2 + 2^3) + 2^4$ $= 15$ [the sum of factors for 8] $+ 16 = 31$.

$32 = 2^5$. Sum of factors: $31 + 2^5 = 31$ [the sum of factors for 16] $+ 32 = 63$.

$64 = 2^6$. Sum of factors: $63 + 2^6 = 63$ [the sum of factors for 32] $+ 64 = 127$.

Answer: 15, 31, 63 and 127

13. Find the number of factors for 630.

Solution: $630 = 2 \times 3 \times 3 \times 5 \times 7 = 2^1 3^2 5^1 7^1$.
Number of factors: $(1 + 1) \times (2 + 1) \times (1 + 1) \times (1 + 1) = 2 \times 3 \times 2 \times 2 = 24$

Answer: 24 (factors)

14. Find the sum of the factors that are not factors of 30 but are factors of 60.

Solution: 30 is a factor of 60 because $60 \div 30 = 2$ R 0. Thus, the sum of all the factors for 60 but not 30 will be the positive difference between the sum of all the factors for 60 and the sum of all the factors of 30.

$60 = 2^2 3^1 5^1$. The sum of factors: $(2^0 + 2^1 + 2^2) \times (3^0 + 3^1) \times (5^0 + 5^1) = 7 \times 4 \times 6 = 168$
$30 = 2^1 3^1 5^1$. The sum of factors: $(2^0 + 2^1) \times (3^0 + 3^1) \times (5^0 + 5^1) = 3 \times 4 \times 6 = 72$

$168 - 72 = 96$.

Answer: 96

15. Find the number of factors for $13^9 31^4$.

Solution: Number of factors: $(9 + 1) \times (4 + 1) = 10 \times 5 = 50$.

Answer: 50 (factors)

16. How many factors do $2^{14}5^{7}11^{5}$ and $3^{8}7^{3}13^{12}$ share?

Solution: The two numbers share no prime factors which means that the only number that they are both divisible by is 1. Thus, they only share 1 factor.

Answer: 1 (factor)

17. Find the number of factors for $11^{3}13^{1}17^{14}$.

Solution: Number of factors: $(3 + 1) \times (1 + 1) \times (14 + 1) = 4 \times 2 \times 15 = 120$.

Answer: 120 (factors)

18. Find the positive difference between the number of factors and the sum of all the factors for 210.

Solution: $210 = 2 \times 3 \times 5 \times 7 = 2^{1}3^{1}5^{1}7^{1}$.
Number of factors: $(1 + 1) \times (1 + 1) \times (1 + 1) \times (1 + 1) = 2 \times 2 \times 2 \times 2 = 16$
Sum of factors: $(2^{0} + 2^{1}) \times (3^{0} + 3^{1}) \times (5^{0} + 5^{1}) \times (7^{0} + 7^{1}) = 3 \times 4 \times 6 \times 8 = 576$

$576 - 16 = 560$

Answer: 560

19. How many factors does 42 and 630 share?

Solution: 42 is a factor of 630 because $630 \div 42 = 15$ R 0. Thus, all factors of 42 will be shared by 630.
$42 = 2 \times 3 \times 7 = 2^{1}3^{1}7^{1}$. Number of factors: $(1 + 1) \times (1 + 1) \times (1 + 1) = 2 \times 2 \times 2 = 8$

Answer: 8 (factors)

20. Find the number of the factors for 12,000.

Solution: $12{,}000 = 2 \times 2 \times 2 \times 2 \times 2 \times 3 \times 5 \times 5 \times 5 = 2^{5}3^{1}5^{3}$.
Number of factors: $(5 + 1) \times (1 + 1) \times (3 + 1) = 6 \times 2 \times 4 = 48$

Answer: 48 (factors)

Review Problems

1. Find the number of even factors for $2^5 3^8 7^4$.

Solution: $5 \times (8 + 1) \times (4 + 1) = 5 \times 9 \times 5 = 225$

Answer: 225 even factors

2. Find the number of odd factors for $2^5 3^8 7^4$.

Solution: $1 \times (8 + 1) \times (4 + 1) = 1 \times 9 \times 5 = 45$

Answer: 45 odd factors

3. How many perfect squares are factors of $2^5 3^8 7^4$?

Solution: $5 \div 2 = 2$ R 1; $8 \div 2 = 4$ R 0; $4 \div 2 = 2$ R 0
$(2 + 1) \times (4 + 1) \times (2 + 1) = 3 \times 5 \times 3 = 45$

Answer: 45 factors

4. How many factors are divisible by 3 for $3^{10} 7^5 13^2$?

Solution: $10 \times (5 + 1) \times (2 + 1) = 10 \times 6 \times 3 = 180$

Answer: 180 factors

5. How many perfect cubes are factors of $2^4 3^9 11^7$?

Solution: $4 \div 3 = 1$ R 1; $9 \div 3 = 3$ R 0; $7 \div 3 = 2$ R 1
$(1 + 1) \times (3 + 1) \times (2 + 1) = 2 \times 4 \times 3 = 24$

Answer: 24 factors

6. How many perfect squares are between 1 and 100, inclusive?

Solution: $1^2, 2^2 \ldots, 10^2 = 1, 4, \ldots, 100$
1^2 to 10^2.
Answer: **10 perfect squares**

7. How many perfect cubes are between 1 and 100, inclusive?

Solution: $1^3, 2^3, 3^3, 4^3, 5^3 = 1, 8, 27, 64, 125$
1^3 to 4^3.

Answer: **4 perfect cubes**

8. Is 64 a perfect square? A perfect cube?

Solution: $64 = 2^6$. 6 is a multiple of 2 and 3.

Answer: **64 is both a perfect square and a perfect cube, or Yes and Yes.**

9. Find the number of factors that are divisible by 6 for $2^2 3^4 5^3$.

Solution: $6 = 2 \times 3 = 2^1 3^1$
$2 \times 4 \times (3 + 1) = 2 \times 4 \times 4 = 32$

Answer: **32 factors**

10. Find the sum of factors that are divisible by 2 but not divisible by 3 for $2^2 3^4 5^3$.

Solution: $(2^1 + 2^2) \times (3^0) \times (5^0 + 5^1 + 5^2 + 5^3) = 6 \times 1 \times 156 = 936$

Answer: **936**

11. Find the number of factors that are perfect squares for $2^2 3^4 5^3$.

Solution: $2 \div 2 = 1$ R 0; $4 \div 2 = 2$ R 0; $3 \div 2 = 1$ R 1
$(1 + 1) \times (2 + 1) \times (1 + 1) = 2 \times 3 \times 2 = 12$

Answer: **12 factors**

12. Find the sum of all the factors that are perfect squares for $2^2 3^3 5^3$.

Solution: $(2^0 + 2^2) \times (3^0 + 3^2) \times (5^0 + 5^2) = 5 \times 10 \times 26 = 1300$

Answer: 1,300

13. Find the number of factors that are perfect cubes for $2^2 3^4 5^3$.

Solution: $2 \div 3 = 0$ R 2; $4 \div 3 = 1$ R 1; $3 \div 3 = 1$ R 0
$(0 + 1) \times (1 + 1) \times (1 + 1) = 1 \times 2 \times 2 = 4$

Answer: 4 factors

14. Find the number of even factors for $2^2 3^4 5^2 7^4$.

Solution: $2 \times (4 + 1) \times (2 + 1) \times (4 + 1) = 2 \times 5 \times 3 \times 5 = 150$

Answer: 150 even factors

15. Find the number of odd factors for $2^2 3^4 5^2 7^4$.

Solution: $1 \times (4 + 1) \times (2 + 1) \times (4 + 1) = 1 \times 5 \times 3 \times 5 = 75$

Answer: 75 odd factors

*16. Find the sum of all the factors that are perfect squares and are divisible by 4 for $2^5 3^3 5^3$

Solution: $(2^2 + 2^4) \times (3^0 + 3^2) \times (5^0 + 5^2) = 20 \times 10 \times 26 = 5200$

Answer: 5,200

17. Find the sum of all the even factors for $2^2 3^2 5^2$.

Solution: $(2^1 + 2^2) \times (3^0 + 3^1 + 3^2) \times (5^0 + 5^1 + 5^2) = 6 \times 13 \times 31 = 2{,}418$

Answer: 2,418

18. Find the sum of all the odd factors for $2^2 3^2 5^2$.

Solution: $(2^0) \times (3^0 + 3^1 + 3^2) \times (5^0 + 5^1 + 5^2) = 1 \times 13 \times 31 = 403$

Answer: 403

*19. Find the number of factors that are divisible by 30 but not divisible by 270 for $2^{20} 3^{15} 5^5$.

Solution: $30 = 2^1 3^1 5^1$; $270 = 2^1 3^3 5^1$
Only 2 powers of 3 will work: 3^1 and 3^2.
$20 \times 2 \times 5 = 200$

Answer: 200 factors

20. Find the number of factors that are divisible by 12 for $2^2 3^4 5^3$.

Solution: $12 = 2^2 3^1$.
$1 \times 4 \times (3 + 1) = 1 \times 4 \times 4 = 16$

Answer: 16 factors

Review Problems

1. Amy bought a vanilla ice cream cone at the county fair for \$6.87 and received \$2.56 in change. How much money did she give the cashier? Express your answer in dollars.

Solution: Amount Given = Price of Item + Change
Amount Given = \$6.87 + \$2.56 = \$9.43

Answer: \$9.43

2. Josh gave the cashier \$1,124.89 to pay for a laptop computer which cost \$1,089.39. How much money should he receive in change? Express your answer in dollars.

Solution: Change = Amount Given – Price of Item
Change = \$1,124.89 – \$1,089.39 = \$35.50

Answer: \$35.50

3. Hannah bought 3 pencils, which cost 28 cents each, with a one-dollar bill. How much did she receive in change? Express your answer in cents.

Solution: Total Cost: 3 × 28¢ = 84¢. \$1.00 = 100¢
Change = Amount Given – Price of Items
Change = 100¢ – 84¢ = 16¢.

Answer: 16¢

4. Sasha has 3 half-dollars, 8 quarters, 11 dimes, 7 nickels and 13 pennies in her pink piggy bank. How much money does she have in total?

Solution:
3 × \$0.50 + 8 × \$0.25 + 11 × \$0.10 + 7 × \$0.05 + 13 × \$0.01
= \$1.50 + \$2.00 + \$1.10 + \$0.35 + \$0.13 = \$5.08

Answer: \$5.08

5. Ross has quarters in 4 stacks with each stack having 8 quarters and dimes in 5 stacks with each stack having 7 dimes. How much money does he have in total? Express your answer in dollars.

Solution: $4 \times 8 \times \$0.25 + 5 \times 7 \times \$0.10 = 32 \times \$0.25 + 35 \times \$0.10 = \$8.00 + \$3.50 = \$11.50$

Answer: $11.50

6. How many ways are there to make 15 cents using pennies and/or nickels?

Solution:

Pennies (1¢)	Nickels (5¢)	Total (15¢)
0	3	15¢
5	2	15¢
10	1	15¢
15	0	15¢

Answer: 4 (ways)

7. The tickets to a concert are priced at $154 for two or $83 for one. How much would Cassie and Jacob save if they bought the tickets for two instead of paying individually for each ticket? Express your answer in dollars.

Solution:
First Option: $154
Second Option: $\$83 \times 2 = \$166$
$\$166 - \$154 = \$12.00$

Answer: $12.00

8. At a convenience store, pens sell for 99 cents each, pencils sell for 49 cents each and erasers sell for 39 cents each. If Riley needs 2 pens, 3 pencils and 5 erasers, how much money does he need? Express your answer in dollars.

Solution: $2 \times \$0.99 + 3 \times \$0.49 + 5 \times \$0.39 = \$1.98 + \$1.47 + \$1.95 = \$5.40$

Answer: $5.40

9. At the candy store, Alex bought 4 chocolate bars costing $2 each and 5 lollipops costing 50 cents each. If Alex had a $20-bill in his wallet, how much money did he have left after purchasing the chocolate bars and lollipops? Express your answer in dollars.

Solution: 4 × $2 + 5 × $0.50 = $8.00 + $2.50 = $10.50
$20.00 – $10.50 = $9.50

Answer: $9.50

10. 3 adults and 2 senior citizens take 4 children to see a movie. If one adult ticket cost $12, one senior citizen ticket cost $10 and one child ticket cost $7.50, how much was the total cost for all the tickets?

Solution: 3 × $12.00 + 2 × $10.00 + 4 × $7.50 = $36.00 + $20.00 + $30.00 = $86.00

Answer: $86.00

11. Rank these deals from least expensive to most expensive for one sunglasses: 1 sunglasses for $14, 3 sunglasses for $40 or 9 sunglasses for $124.

Solution: For each deal, find the price for 1 sunglasses.
Deal 1: 14 dollars for 1 sunglasses (#3)
Deal 2: $\frac{40}{3} \approx 13.33$ dollars for 1 sunglasses (#1)
Deal 3: $\frac{124}{9} \approx 13.77$ dollars for 1 sunglasses (#2)

Answer: 1) 3 sunglasses for $40; 2) 9 sunglasses for $124; 3) 1 sunglasses for $14

12. A store sells boxes with two options: 1 for $3.00 or 11 for $30.00. What is the least amount that John needs to spend to get exactly 100 boxes?

Solution: Buying 11 boxes individually is 11 × $3.00 = $33.00 which is more expensive than 11 boxes for $30.00. Thus, John should buy as many boxes with the second option.

100 ÷ 11 = 9 R 1.
1 × $3.00 + 9 × $30.00 = $3.00 + $270.00 = $273.00

Answer: $273.00

13. How many ways are there to make 17 cents using pennies and/or nickels?

Solution:

Pennies (1¢)	Nickels (5¢)	Total (17¢)
2	3	15¢
7	2	17¢
12	1	17¢
17	0	17¢

Answer: 4 (ways)

14. Julie takes a taxi from her house to the airport. The taxi charges $4.50 to start the meter and $1.10 for each mile traveled. If the distance is 15 miles from Julie's house to the airport, how much will she have to pay the taxi driver when she arrives at the airport? Express your answer in dollars.

Solution: $4.50 + 15 × $1.10 = $4.50 + $16.50 = $21.00

Answer: $21.00

15. A company sells shirts with three different options: 2 for $19, 5 for $45 or 20 for $170. If Logan needs to purchase exactly 132 shirts, what is the least amount he has to pay?

Solution:

Option 1: $\frac{19}{2} = 9.50 \text{ dollars}$ for 1 shirt.

Option 2: $\frac{45}{5} = 9 \text{ dollars}$ for 1 shirt.

Option 3: $\frac{170}{20} = 8.50 \text{ dollars}$ for 1 shirt.

Option 3 is the best priced option followed by Option 2 then Option 1.

132 ÷ 20 = 6 R 12. Option 3 should be used 6 times.
12 ÷ 5 = 2 R 2. Option 2 should be used 2 times.
2 ÷ 2 = 1 R 0. Option 1 should be used 1 time.

6 × \$170 + 2 × \$45 + 1 × \$19 = \$1020 + \$90 + \$19 = \$1,129.00

Answer: \$1,129.00

16. How many ways are there to make 25 cents using quarters, dimes and/or nickels?

Solution:

Nickels (5¢)	Dimes (10¢)	Quarters (25¢)	Total (25¢)
0	0	1	25¢
1	2	0	25¢
3	1	0	25¢
5	0	0	25¢

Answer: 4 (ways)

17. A phone company charges \$15 a month for their services and charges an extra 50 cents per minute for long distance calls. If Josie's July bill is \$99, how many minutes did she call long distance?

Solution: \$15.00 + M × \$0.50 = \$99.00. Subtracting \$15.00 from both sides:
M × \$0.50 = \$99.00 – \$15.00 = \$84.00. Dividing both sides by \$0.50
M = \$84.00 ÷ \$0.50 = 168

Answer: 168 (minutes)

18. 4 adults and 2 children went to visit an aquarium. If one adult ticket cost \$14 and one child ticket cost \$5, how much was the total cost for all the tickets?

Solution: 4 × \$14.00 + 2 × \$5.00 = \$56.00 + \$10.00 = \$66.00

Answer: \$66.00

19. Ten adults and ten children go to see a movie at the theatre. If the total amount came out to be $210 and an adult ticket cost $14, how much did a child ticket cost?

Solution: $10 \times \$14.00 + 10 \times C = \$140 + 10C = \$210$. Subtracting both sides by $140:
$10C = \$210 - \$140 = \$70$. Dividing both sides by 10:
$C = \$70 \div 10 = \$7.00$

Answer: $7.00

*20. How many ways are there to make $20 using $10-bills, $5-bills and/or $1-bills?

Solution:

$1 bill	$5 bill	$10 bill	Total ($20)
0	0	2	$20
0	2	1	$20
5	1	1	$20
10	0	1	$20
0	4	0	$20
5	3	0	$20
10	2	0	$20
15	1	0	$20
20	0	0	$20

Answer: 9 (ways)

Review Problems

1. Simplify $5^{13} \times 25^{7}$ as a power of 5.

Solution: $5^{13} \times (5^{2})^{7} = 5^{13} \times 5^{14} = 5^{(13+14)} = 5^{27}$

Answer: $\mathbf{5^{27}}$

2. Simplify $7^{6} \div 7^{(-4)} \div 7^{10}$.

Solution: $7^{6} \times 7^{4} \times 7^{(-10)} = 7^{(6+4-10)} = 7^{0} = 1$

Answer: **1**

3. Simplify $\frac{1}{8^{-2}}$.

Solution: $\frac{1}{8^{-2}} = 8^{2} = 64$

Answer: **64**

4. Simplify $\sqrt{100}$.

Solution: $\sqrt{100} = \sqrt{2^{2}5^{2}} = 2^{\frac{2}{2}}5^{\frac{2}{2}} = 2^{1}5^{1} = 10$

Answer: **10**

5. Simplify $\sqrt[3]{72}$.

Solution: $\sqrt[3]{72} = \sqrt[3]{2^{3}3^{2}} = 2^{\frac{3}{3}}3^{\frac{2}{3}} = 2^{1}\sqrt[3]{3^{2}} = 2\sqrt[3]{9}$

Answer: $2\sqrt[3]{9}$

6. Simplify $\sqrt{2^{32}} * \sqrt{2^{-32}}$.

Solution: $\sqrt{2^{32}} * \sqrt{2^{-32}} = 2^{\frac{32}{2}} * 2^{\frac{-32}{2}} = 2^{16}2^{-16} = 2^{16-16} = 2^0 = 1$

Answer: 1

7. Simplify $(\sqrt[7]{384})^0$.

Solution: $(\sqrt[7]{384})^0 = 1$

Answer: 1

8. Find the integer portion of the decimal representation of $\sqrt{19}$.

Solution: $4 < \sqrt{19} < 5;$ $\sqrt{16} < \sqrt{19} < \sqrt{25}$

Answer: 4

9. Find the two consecutive integers that are closest to $\sqrt[3]{150}$.

Solution: $5 < \sqrt[3]{150} < 6;$ $\sqrt[3]{125} < \sqrt[3]{150} < \sqrt[3]{216}$

Answer: 5 and 6

10. Simplify $\sqrt[14]{81^7}$.

Solution: $\sqrt[14]{81^7} = \sqrt[14]{(3^4)^7} = \sqrt[14]{3^{28}} = 3^{\frac{28}{14}} = 3^2 = 9$

Answer: 9

11. Simplify $\sqrt[6]{2^{14}} * \sqrt[3]{2^5}$.

Solution: $\sqrt[6]{2^{14}} * \sqrt[3]{2^5} = 2^{\frac{14}{6}} * 2^{\frac{5}{3}} = 2^{\frac{7}{3}+\frac{5}{3}} = 2^4 = 16$

Answer: 16

12. Simplify $\sqrt[3]{2^2 11^1} * \sqrt[6]{2^2 11^4}$.

Solution: $\sqrt[3]{2^2 11^1} * \sqrt[6]{2^2 11^4} = 2^{\frac{2}{3}} 11^{\frac{1}{3}} * 2^{\frac{2}{6}} 11^{\frac{4}{6}} = 2^{\frac{2}{3}+\frac{1}{3}} 11^{\frac{1}{3}+\frac{2}{3}} = 2^1 11^1 = 22$

Answer: 22

13. Simplify $\sqrt[11]{\sqrt[3]{(8^4)^7 * 32^3}}$.

Solution: $\sqrt[11]{\sqrt[3]{(8^4)^7 * 32^3}} = \sqrt[11]{\sqrt[3]{((2^3)^4)^7 * (2^5)^3}} = \sqrt[11]{\sqrt[3]{(2^{12})^7 * 2^{15}}}$
$= \sqrt[11]{\sqrt[3]{2^{84} * 2^{15}}} = \sqrt[11]{\sqrt[3]{2^{99}}} = ((2^{99})^{\frac{1}{3}})^{\frac{1}{11}} = (2^{33})^{\frac{1}{11}} = 2^3 = 8$

Answer: 8

14. Simplify $(\sqrt[3]{27000000})^2$

Solution: $(\sqrt[3]{27000000})^2 = (\sqrt[3]{2^6 3^3 5^6})^2 = (2^{\frac{6}{3}} 3^{\frac{3}{3}} 5^{\frac{6}{3}})^2 = (2^2 3^1 5^2)^2 = 2^4 3^2 5^4$

Answer: $2^4 3^2 5^4$ or 90,000

15. Simplify $\sqrt[3]{\sqrt[3]{\sqrt[2]{9^9}}}$.

Solution:

$\sqrt[3]{\sqrt[3]{\sqrt[2]{9^9}}} = \sqrt[3]{\sqrt[3]{\sqrt[2]{(3^2)^9}}} = \sqrt[3]{\sqrt[3]{\sqrt[2]{3^{18}}}} = ((((3^{18})^{\frac{1}{2}})^{\frac{1}{3}})^{\frac{1}{3}}) = 3^{18*\frac{1}{2}*\frac{1}{3}*\frac{1}{3}} =$
$3^1 = 3$

Answer: 3

16. Find the largest perfect square less than 500.

Solution: $\sqrt{500} = 10\sqrt{5} \approx 10 * 2.236 = 22.36$
Therefore, $22^2 = 484$ is the largest perfect square less than 500

Answer: $\mathbf{22^2}$ **or 484**

17. How many perfect squares are between 100 and 1,000, inclusive?

Solution: $\sqrt{100} = 10, \sqrt{1000} = 10\sqrt{10} \approx 10 * 3.162 = 31.62$
$10^2, 11^2, \ldots, 31^2$
Using the counting number inclusive formula: $(b - a) + 1 = (31 - 10) + 1 = 22$

Answer: 22 perfect squares

18. Find the largest perfect square less than 9,000.

Solution: $\sqrt{9000} = 30\sqrt{10} \approx 30 * 3.162 = 94.86$
Therefore, $94^2 = 8{,}836$ is the largest perfect square less than 9,000

Answer: $\mathbf{94^2}$ **or 8,836**

19. Simplify $\sqrt[2]{11} * \sqrt[3]{11} * \sqrt[6]{11}$.

Solution: $\sqrt[2]{11} * \sqrt[3]{11} * \sqrt[6]{11} = 11^{\frac{1}{2}} * 11^{\frac{1}{3}} * 11^{\frac{1}{6}} = 11^{\frac{1}{2}+\frac{1}{3}+\frac{1}{6}} = 11^1 = 11$

Answer: 11

20. Simplify $\sqrt[9]{5^2} * \sqrt[18]{5^2} * \sqrt[3]{5^8}$.

Solution: $\sqrt[9]{5^2} * \sqrt[18]{5^2} * \sqrt[3]{5^8} = 5^{\frac{2}{9}} * 5^{\frac{2}{18}} * 5^{\frac{8}{3}} = 5^{\frac{2}{9}+\frac{2}{18}+\frac{8}{3}} = 5^3 = 125$

Answer: 125

Review Problems

1. Express $\frac{15}{20}$ as a reduced fraction.

Solution: $\frac{15}{20} = \frac{3^1 5^1}{2^2 5^1} = \frac{3^1}{2^2} = \frac{3}{4}$

Answer: $\frac{3}{4}$

2. Solve: ½ + ⅓ + ¼. Express your answer as a reduced mixed number.

Solution:

$$\frac{1}{2}*\frac{3}{3}*\frac{4}{4}+\frac{1}{3}*\frac{2}{2}*\frac{4}{4}+\frac{1}{4}*\frac{2}{2}*\frac{3}{3} = \frac{12}{24}+\frac{8}{24}+\frac{6}{24} = \frac{12+8+6}{24} = \frac{26}{24} = \frac{13}{12} = 1\frac{1}{12}$$

Answer: $1\frac{1}{12}$

3. Convert $\frac{5}{4}$ to a decimal and a percent.

Solution: $\frac{5}{4} = 1\frac{1}{4} = 1.25$

$1.25 * 100 = 125 \text{ percent}$ = 125%

Answer: 1.25; 125%

4. Solve: $\frac{100}{120} * \frac{4}{14} * \frac{35}{8}$. Express your answer as an improper fraction.

Solution: $\frac{2^2 5^2 * 2^2 * 5^1 7^1}{2^3 3^1 5^1 * 2^1 7^1 * 2^3} = \frac{2^4 5^3 7^1}{2^7 3^1 5^1 7^1} = \frac{5^2}{2^3 3^1} = \frac{25}{24}$

Answer: $\frac{25}{24}$

5. What percent of 100 is 35?

Solution: $\frac{35}{100} * 100 = 35 \text{ percent}$ = 35%

Answer: 35%

6. What percent of 25 is 6?

Solution: $\frac{6}{25} * 100 = 24 \text{ percent}$ = 24%

Answer: 24%

7. Solve: $\frac{13}{12} \div \frac{39}{48}$. Express your answer as a reduced mixed number.

Solution: $\frac{13}{12} * \frac{48}{39} = \frac{4}{3} = 1\frac{1}{3}$

Answer: $1\frac{1}{3}$

8. What percent of 40 is 28?

Solution: $\frac{28}{40} * 100 = \frac{7}{10} * 100 = 70 \text{ percent}$ = 70%

Answer: 70%

9. Herald answered 80% of the 120 questions on his test correctly. How many questions did he get wrong?

Solution: Wrong = (100 – 80)% = 20%

$\frac{20}{100} * 120 = \frac{1}{5} * 120 = 24$

Answer: 24 questions

10. Arnold, Bob and Carl are sharing a pizza. Arnold ate ⅙ of the pizza, Bob ate ¼ of the pizza and Carl ate the rest. What fraction of the pizza did Carl eat? Express your answer as a reduced proper fraction.

Solution:

$$1-\frac{1}{6}-\frac{1}{4}=1*\frac{6}{6}*\frac{4}{4}-\frac{1}{6}*\frac{4}{4}-\frac{1}{4}*\frac{6}{6}=\frac{24}{24}-\frac{4}{24}-\frac{6}{24}=\frac{24-4-6}{24}=\frac{14}{24}=\frac{7}{12}$$

Answer: $\frac{7}{12}$

*11. If there are 15 boys and 25 girls in a book club, what percent of the book club members are boys?

Solution: $\frac{15}{15+25}*100=\frac{15}{40}*100=\frac{3}{8}*100=37.5 \text{ percent}$ = 37.5%

Answer: 37.5%

12. 800 students attend a high school. If the enrollment increases by 20% the next year, how many students will attend the high school next year?

Solution: $800*\frac{(100+20)}{100}=800*\frac{120}{100}=8*120=960$

Answer: 960 students

13. If there are 8 children and 12 adults at the park, what percent of the people at the park are adults?

Solution: $\frac{12}{8+12}*100=\frac{12}{20}*100=12*5=60 \text{ percent}$ = 60%

Answer: 60%

14. Sandra bought five shirts at a clothing store. If the price of each shirt was $25 and she used a 40% off coupon that could only be used on one shirt, how much did she pay for the five shirts?

Solution: $25 * \frac{(100-40)}{100} = 25 * \frac{60}{100} = 25 * \frac{3}{5} = 15$

$15 + (5 – 1) × $25 = $15 + $100 = $115

Answer: $115.00

15. A smartphone is selling for $800 in January. One month later, in February, the phone was 20% off the original price. In the following month, March, the phone's price increased by 20% from February. How much was the phone's price in March?

Solution: $800 * \frac{(100-20)}{100} * \frac{(100+20)}{100} = 800 * \frac{4}{5} * \frac{6}{5} = 768$

Answer: $768.00

16. Barbara has a plastic container that consists of red, blue and green marbles. If there are 200 marbles in the container in which 25% are green, 20% are red and the remaining marbles are blue, how many marbles are blue?

Solution: Blue: (100 – 25 – 20)% = 55%

$200 * \frac{55}{100} = 110$

Answer: 110 marbles

17. There are 64 students consisting of 36 boys and 28 girls in Mr. Smith's math class. If ¼ of the class is absent one day, and ⅓ of the boys are absent, how many girls are absent that day?

Solution: $\text{Total Absent:} \ = 64 * \frac{1}{4} = 16$

$\text{Boys Absent:} \ = 36 * \frac{1}{3} = 12$

16 – 12 = 4 girls

Answer: 4 girls

18. On an exam with 275 problems, Ruth answered 76% of the questions correct, left 4% of the questions blank and got the remaining questions incorrect. If she receives 1 point for every correct answer, 0 points for every blank answer and loses ⅕ of a point for every incorrect answer, what was her score on the test?

Solution: Correct: $275 * \frac{76}{100} = 275 * \frac{19}{25} = 209$

Blank: $275 * \frac{4}{100} = 275 * \frac{1}{25} = 11$

Wrong: $275 - 209 - 11 = 55$

Score: $(209 * 1) + (11 * 0) + (55 * -\frac{1}{5}) = 209 + 0 - 11 = 198$

Answer: 198 points

*19. A laptop's price increased by 25%. What percent must the laptop's new price be reduced in order to get back to the original price?

Solution: $x * \frac{(100+25)}{100} = 1$. Dividing both sides to get x:

$x = 1 \div \frac{(100+25)}{100} = 1 * \frac{100}{(100+25)} = \frac{100}{125} = \frac{4}{5}$

$\frac{4}{5} * 100 = 80 \text{ percent}$

(100 – 80)% = 20%

Answer: 20%

20. Joshua has 96 jellybeans. He gives ⅓ of the jellybeans to his friend, Paul. Then he gives ¼ of the jellybeans he has left to his brother, George. And finally, he gives ⅙ of the remaining jellybeans to his sister, Nancy. How many jellybeans does Joshua have left at the end?

Solution: $96 * (1 - \frac{1}{3}) * (1 - \frac{1}{4}) * (1 - \frac{1}{6}) = 96 * \frac{2}{3} * \frac{3}{4} * \frac{5}{6} = 40$

Answer: 40 jellybeans

Review Problems

1. The odds of Jack losing to Jill at a game is 13 to 7. What is the probability that Jack will win? Express your answer as a percent.

Solution: Failures to Successes = 13 to 7

$$\text{P(Success)} = \frac{7}{13+7} = \frac{7}{20}$$

$$\frac{7}{20} * 100 = 35 \text{ percent}$$ = 35%

Answer: 35%

2. A bag has red and green marbles and the ratio of red to green is 3 to 8. If there are 33 red marbles, how many green marbles are there?

Solution: 33 (marbles) ÷ 3 (parts) = 11 (factor)
8 (parts) × 11 (factor) = 88 green marbles

Answer: 88 green marbles

3. There are 40 people in a room where the ratio of men to women is 3:5. What is the positive difference between the number of men and number of women in the room?

Solution: Total Parts: 3 + 5 = 8
40 (people) ÷ 8 (parts) = 5 (factor)
Men = 5 × 3 = 15; Women = 5 × 5 = 25
Women – Men = 25 – 15 = 10

Answer: 10

4. A bag has yellow and purple marbles with a ratio of 1 to 1. There is a total of 18 marbles. If 3 more yellow marbles were added, what is the new ratio of yellow to purple marbles?

Solution: Total Parts: 1 + 1 = 2
18 ÷ 2 = 9 (factor)
Yellow = 9 × 1 = 9; Purple = 9 × 1 = 9
New Yellow = 9 + 3 = 12

Yellow to Purple = 12 to 9 = $\frac{12}{9} = \frac{4}{3}$ = 4 to 3

Answer: 4 to 3 or 4:3

5. The ratio of boys to girls in a fifth-grade class is 9 to 8. If there are 72 girls, how many total fifth-graders are there?

Solution: 72 ÷ 8 = 9 (factor)
Total Parts: 9 + 8 = 17
17 × 9 = 153

Answer: 153 fifth-graders

6. There are apples, oranges and bananas in a large basket. If the ratio of apples to oranges to bananas is 2:5:3 and the total number of fruits is 170, how many apples are there?

Solution: Total Parts: 2 + 5 + 3 = 10
170 ÷ 10 = 17 (factor)
2 × 17 = 34 apples

Answer: 34 apples

7. The odds that Daniel draws a red sock from a drawer with red, yellow and blue socks is 7 to 10. If it is equally likely to draw a yellow sock as it is to draw a blue sock, what fraction of the total number of socks is blue?

Solution: $\mathrm{P(red)} = \frac{7}{10+7} = \frac{7}{17}$

$$\mathrm{P(yellow)} = \mathrm{P(blue)} = \frac{1 - \mathrm{P(red)}}{2} = \frac{1 - \frac{7}{17}}{2} = \frac{\frac{10}{17}}{2} = \frac{5}{17}$$

Answer: $\frac{5}{17}$

8. Jacob has nickels, dimes and quarters. The ratio of nickels to dimes to quarters is 11 to 7 to 6. If he has 72 total in nickels and dimes, how many quarters does he have?

Solution: Total Parts for Nickels and Dimes: $11 + 7 = 18$
$72 \div 18 = 4$ (factor)
$4 \times 6 = 24$ quarters

Answer: 24 quarters

9. The ratio of the side lengths of a quadrilateral is 8:6:4:6. If the perimeter is 264, find the positive difference between the longest and shortest side lengths.

Solution: Total Parts: $8 + 6 + 4 + 6 = 24$
$264 \div 24 = 11$ (factor)
Shortest: $11 \times 4 = 44$; Longest: $11 \times 8 = 88$
$88 - 44 = 44$

Answer: 44

10. In a parking lot, the ratio of the number of cars to the number of trucks is 1 to 2. If there are 99 more trucks than cars, how many total vehicles are in the parking lot?

Solution: Difference in parts for trucks and car: $2 - 1 = 1$
$99 \div 1 = 99$ (factor)
Total Parts: $1 + 2 = 3$
$99 \times 3 = 297$ vehicles

Answer: 297 vehicles

11. The ratio of unicycles to bicycles to tricycles in a park is 3:2:1. If there are 20 more unicycles than tricycles, how many bicycles are there?

Solution: Difference in parts for unicycles and tricycles: $3 - 1 = 2$
$20 \div 2 = 10$ (factor)
$10 \times 2 = 20$ bicycles

Answer: 20 bicycles

12. The ratio of DVD's to books on a shelf is 7 to 5. If there are 60 books, how many DVD's need to be removed from the shelf to give a new ratio of DVD's to books to be 2 to 3?

Solution: 60 ÷ 5 = 12 (factor)
DVD's = 12 × 7 = 84

If books stay the same, 60 ÷ 3 = 20 (factor)
20 × 2 = 40 DVD's (after)
84 – 40 = 44

Answer: 44 DVD's

13. A room has pink and green balls. The ratio of pink to green balls is 29 to 14. If 195 pink balls were removed from the room, the new ratio would be 1 to 1. How many green balls are in the room?

Solution: Difference in parts for pink and green: 29 – 14 = 15
195 ÷ 15 = 13 (factor)
Green = 13 × 14 = 182

Answer: 182 green balls

*14. The ratio of boys to girls in a class is 5 to 4. If the number of girls were doubled, there would be 27 more girls than boys. What is the sum of the the number of boys and girls before the girls were doubled?

Solution: The number of parts after doubling: 5 to (4 × 2) = 5 to 8
Difference in parts: 8 – 5 = 3
27 ÷ 3 = 9 (factor)
Total Parts (before): 5 + 4 = 9
9 × 9 = 81 students

Answer: 81 students

15. In a quadrilateral, the ratio of the angles is 4:5:11:4. Find the measure of the largest angle. (Hint: The sum of the angles for a quadrilateral is 360 degrees.)

Solution: Sum of angles: (n – 2) × 180 = (4 – 2) × 180 = 2 × 180 = 360
Total Parts: 4 + 5 + 11 + 4 = 24
360 ÷ 24 = 15 (factor)
15 × 11 = 165 degrees

Answer: 165 degrees

16. There are squares, triangles and circles drawn on the driveway. The ratio of the number of squares to the number of triangles to the number of circles is 14:9:10. If there are 12 more squares than circles, how many total shapes are drawn on the driveway?

Solution: Difference in parts for squares and circles: 14 – 10 = 4
12 ÷ 4 = 3 (factor)
Total Parts: 14 + 9 + 10 = 33
3 × 33 = 99 shapes

Answer: 99 shapes

17. If the ratio of teachers to boys to girls is 2 to 8 to 7, what fraction of the school population is students?

Solution: $\frac{8+7}{2+8+7} = \frac{15}{17}$

Answer: $\frac{15}{17}$

18. The ratio of dogs to rabbits to cats at an animal shelter is 19 to 15 to 17. If 32 dogs and 16 cats are adopted, the new ratio of dogs to rabbits to cats would be 1 to 1 to 1. How many total animals are at the shelter after the adoptions?

Solution: Difference in parts for dogs and rabbits: $19 - 15 = 4$
$32 \div 4 = 8$ (factor)
Dogs (before): $8 \times 19 = 152$; Dogs (after): $152 - 32 = 120$
Rabbits: $8 \times 15 = 120$
Cats (before): $8 \times 17 = 136$; Cats (after): $136 - 16 = 120$
Total (after): $120 + 120 + 120 = 360$

Answer: 360 animals

19. If the ratio of teachers to boys to girls is 3 to 11 to 15 and there are 492 more girls than teachers, how many boys are there?

Solution: Difference in parts for girls and teachers: $15 - 3 = 12$
$492 \div 12 = 41$ (factor)
Boys $= 41 \times 11 = 451$

Answer: 451 boys

*20. A bag has orange and green marbles with a ratio of orange to green equal to 9 to 5. If 71 orange marbles are removed, there would still be 5 more orange marbles than green marbles. How many marbles are there before the orange marbles were removed from the bag?

Solution: $71 + 5 = 76$ more oranges than green
Difference in parts: $9 - 5 = 4$
$76 \div 4 = 19$ (factor)
Total Parts: $9 + 5 = 14$
$19 \times 14 = 266$

Answer: 266 marbles

Review Problems

1. Given an arithmetic sequence, 2, 5, 8, 11, …, find the 9th term.

Solution: $a_1 = 2, d = 5 - 2 = 3$
$a_9 = a_1 + d(9-1) = 2 + 3 * 8 = 2 + 24 = 26$

Answer: 26

2. Given an arithmetic sequence, -20, -15, -10, -5, …, find the 71st term.

Solution: $a_1 = -20, d = -15 - (-20) = 5$
$a_{71} = a_1 + d(71-1) = -20 + 5 * 70 = -20 + 350 = 330$

Answer: 330

3. Find the sum of the first 10 terms of the sequence: 2, 7, 12, 17, …

Solution: $a_1 = 2, d = 7 - 2 = 5$
$a_{10} = a_1 + d(10-1) = 2 + 5 * 9 = 2 + 45 = 47$

$$S_{10} = \frac{(a_1 + a_{10}) * 10}{2} = \frac{(2+47) * 10}{2} = \frac{490}{2} = 245$$

Answer: 245

4. Find the sum of the first 15 terms of the sequence: $-14, -12, -10, -8, \ldots$

Solution: $a_1 = -14, d = -12 - (-14) = 2$
$a_{15} = -14 + 2(15-1) = -14 + 28 = 14$

$$S_{15} = \frac{(-14+14) * 15}{2} = \frac{0 * 15}{2} = 0$$

Answer: 0

5. Find the missing terms in the sequence: ___, 11, ___, 29, 38, ___

Solution: $d = a_5 - a_4 = 38 - 29 = 9$
Each consecutive term is separated by the common difference.
$a_1 = a_2 - d = 11 - 9 = 2$
$a_3 = a_2 + d = 11 + 9 = 20$ or $a_3 = a_4 - d = 29 - 9 = 20$
$a_6 = a_5 + d = 38 + 9 = 47$

Answer: 2, 20, and 47

6. Find the missing terms in the sequence: –55, ___, 215, ___, ___, 620, 755

Solution: $d = a_7 - a_6 = 755 - 620 = 135$
$a_2 = a_1 + d = -55 + 135 = 80$ or $a_2 = a_3 - d = 215 - 135 = 80$
$a_4 = a_3 + d = 215 + 135 = 350$
$a_5 = a_6 - d = 620 - 135 = 485$

Answer: 80, 350 and 485

7. What is the next term in the sequence? 80, 71, 62, 53,…

Solution: $d = 71 - 80 = -9$
$a_5 = a_4 + d = 53 + (-9) = 44$

Answer: 44

8. If James' allowance is $15 dollars in January, $35 in February, $55 in March and so on with every monthly allowance increasing by $20, how much money would he have received in allowance in a year?

Solution: $a_1 = 15, d = 20$

There are twelve months in a year: $a_{12} = 15 + 20(12 - 1) = 15 + 20 * 11 = 235$

$$S_{12} = \frac{(15 + 235) * 12}{2} = \frac{250 * 12}{2} = 250 * 6 = 1500$$

Answer: $1500

9. Find the sum of the first 15 positive integers.

Solution: $a_1 = 1, a_{15} = 15$

$$S_{15} = \frac{(1+15)*15}{2} = \frac{16*15}{2} = 8*15 = 120$$

Answer: 120

10. Find the missing terms in the sequence: 88, ___, 112, ___, 136, ___

Solution: From the first term to the second term, they are one common difference apart. Therefore, from the first term to the third term, they are two common differences apart.

$a_3 - a_1 = 112 - 88 = 24 = 2d$

If $2d = 24$, then $d = 24 \div 2 = 12$.

$a_2 = a_1 + d = 88 + 12 = 100$

$a_4 = a_3 + d = 112 + 12 = 124$

$a_6 = a_5 + d = 136 + 12 = 148$

Answer: 100, 124, and 148

11. If a grandfather clock chimes once at 1 o'clock, twice at 2 o'clock, thrice at 3 o'clock and so on for each of the 12 hours, how many times does it chime in total over the 12 hour period?

Solution: $a_1 = 1, a_{12} = 12$

$$S_{12} = \frac{(1+12)*12}{2} = \frac{13*12}{2} = 13*6 = 78$$

Answer: 78 (chimes)

12. What is the product of the 10th term and the 100th term of the following sequence: −297, −294, −291, … ?

Solution: $a_1 = -297, d = -294 - (-297) = 3$

$a_{10} = -297 + 3*(10-1) = -297 + 27 = -270$

$a_{100} = -297 + 3*(100-1) = -297 + 297 = 0$

$a_{10} * a_{100} = -270 * 0 = 0$

Answer: 0

13. Find the missing terms, then find the sum of the first 20 terms for the following sequence: 9, ___, 25, 33, ___, ...

Solution: $d = 33 - 25 = 8$

$a_2 = a_1 + d = 9 + 8 = 17$

$a_5 = a_4 + d = 33 + 8 = 41$

$a_{20} = 9 + 8 * (20 - 1) = 9 + 152 = 161$

$$S_{20} = \frac{(9 + 161) * 20}{2} = \frac{170 * 20}{2} = 170 * 10 = 1700$$

Answer: 17 and 41; 1700

14. What is the sum of the terms between the 10th term and the 30th term, inclusive, of the following sequence: 1, 16, 31, 46, ... ?

Solution: There are (30 – 10) + 1 = 21 numbers between 10 and 30, inclusive.

$a_1 = 1, d = 16 - 1 = 15$

$a_{10} = 1 + 15 * (10 - 1) = 136$

$a_{30} = 1 + 15 * (30 - 1) = 436$

$$S = \frac{(136 + 436) * 21}{2} = \frac{572 * 21}{2} = 286 * 21 = 6006$$

Answer: 6,006

15. Find the sum of the first 25 positive even integers.

Solution: {2, 4, 6, 8, 10, ... } $a_1 = 2, a_2 = 4, d = 4 - 2 = 2$

$a_{25} = 2 + 2 * (25 - 1) = 50$

$$S_{25} = \frac{(2 + 50) * 25}{2} = \frac{52 * 25}{2} = 26 * 25 = 650$$

Answer: 650

16. Find the positive difference between the sum of the first 30 positive even integers and the sum of the first 30 positive odd integers.

Solution: {2, 4, 6, 8, 10, … } $a_1 = 2, d = 2, a_{30} = 2 + 2 * (30 - 1) = 60$

{1, 3, 5, 7, 9, … } $b_1 = 1, d = 3 - 1 = 2, b_{30} = 1 + 2 * (30 - 1) = 59$

$$S_a = \frac{(2 + 60) * 30}{2} = \frac{62 * 30}{2} = 31 * 30 = 930$$

$$S_b = \frac{(1 + 59) * 30}{2} = \frac{60 * 30}{2} = 30 * 30 = 900$$

$930 - 900 = 30$

Faster Way: Notice that each of the even terms is 1 more than each of the odd terms. For example, the first terms are 2 vs. 1 and the second terms are 4 vs. 3. Since there are 30 terms, the difference is 30.

Answer: 30

17. A store has a stack of soda cans. The top row has 1 can, the second row has 8 cans, and so on in an arithmetic sequence until the final row with 50 cans. How many cans in total are in the stack?

Solution: {1, 8, …, 50} $a_1 = 1, d = 8 - 1 = 7$

$a_n = 1 + 7 * (n - 1) = 50$. Subtract 1 from both sides:

$7 * (n - 1) = 49$. Divide both sides by 7:

$(n - 1) = 7$. Add one to both sides:

$n = 8$. There are 8 terms.

$$S_8 = \frac{(1 + 50) * 8}{2} = \frac{51 * 8}{2} = 51 * 4 = 204$$

Answer: 204 (cans)

18. Find the missing terms: 11, ___, ___, ___, 251, ___

Solution: $a_1 = 11, a_5 = 251$. The number of common differences that separates the first and fifth term is 4.

Thus, $a_5 - a_1 = 251 - 11 = 240 = 4d$. Dividing both sides by 4:
$d = 240 \div 4 = 60$

$a_2 = a_1 + d = 11 + 60 = 71$
$a_3 = a_2 + d = 71 + 60 = 131$
$a_4 = a_3 + d = 131 + 60 = 191$
$a_6 = a_5 + d = 251 + 60 = 311$

Answer: 71, 131, 191, and 311

*19. Find the difference between the sum of the 1st and 50th term and the sum of the 2nd and 49th term of the sequence: –146, –96, –46, …

Solution: $a_1 = -146, a_2 = -96, d = -96 - (-146) = 50$
$a_{50} = -146 + 50 * (50 - 1) = 2304$
$a_{49} = -146 + 50 * (49 - 1) = 2254$
$a_1 + a_{50} = -146 + 2304 = 2158$
$a_2 + a_{49} = -96 + 2254 = 2158$
$2158 - 2158 = 0$

Faster Way: The trick is that the sums will be the same because of the common difference. The 2nd term is 50 more than the 1st term but the 49th term will be 50 less than the 50th term. The two sums will cancel out and give a difference of 0.

Answer: 0

*20. If the first term of a sequence is 3, the last term is 39 and the sum of the terms from first to last is 210, what is the product of the common difference and the number of terms?

Solution: $S_n = \frac{(3 + 39) * n}{2} = \frac{42n}{2} = 21n = 210$
Dividing both sides by 21: $n = 210 \div 21 = 10$. There are 10 terms.
$a_1 = 3, a_{10} = 39, d = ?$
$a_{10} = 3 + d * (10 - 1) = 3 + 9d = 39$. Subtracting both sides by 3:
$9d = 39 - 3 = 36$. Dividing both sides by 9. $d = 36 \div 9 = 4$
$n * d = 10 * 4 = 40$

Answer: 40

Review Problems

1. Convert 5 pounds to ounces.

Solution: $$\frac{5 \text{ pounds}}{1} * \frac{16 \text{ ounces}}{1 \text{ pound}} = \frac{5}{1} * \frac{16 \text{ ounces}}{1} = 80 \text{ ounces}$$

Answer: 80 (ounces)

2. How many hours are in 7 days?

Solution: $$\frac{7 \text{ days}}{1} * \frac{24 \text{ hours}}{1 \text{ day}} = \frac{7}{1} * \frac{24 \text{ hours}}{1} = 168 \text{ hours}$$

Answer: 168 (hours)

3. How many yards are in a mile?

Solution: $$\frac{1 \text{ mile}}{1} * \frac{5280 \text{ feet}}{1 \text{ mile}} * \frac{1 \text{ yard}}{3 \text{ feet}} = \frac{1}{1} * \frac{5280}{1} * \frac{1 \text{ yard}}{3} = 1760 \text{ yards}$$

Answer: 1760 (yards)

4. How many seconds are in 30 minutes?

Solution: $$\frac{30 \text{ minutes}}{1} * \frac{60 \text{ seconds}}{1 \text{ minute}} = \frac{30}{1} * \frac{60 \text{ seconds}}{1} = 1800 \text{ seconds}$$

Answer: 1800 (seconds)

5. What is 35 feet in inches?

Solution: $$\frac{35 \text{ feet}}{1} * \frac{12 \text{ inches}}{1 \text{ foot}} = \frac{35}{1} * \frac{12 \text{ inches}}{1} = 420 \text{ inches}$$

Answer: 420 (inches)

6. An action-figure weighs 8 pounds and 4 ounces. How much does it weigh in ounces?

Solution: First, convert 8 pounds to ounces. Then add this to 4 ounces.

$$\frac{8 \text{ pounds}}{1} * \frac{16 \text{ ounces}}{1 \text{ pound}} = \frac{8}{1} * \frac{16 \text{ ounces}}{1} = 128 \text{ ounces}$$

128 ounces + 4 ounces = 132 ounces

Answer: 132 (ounces)

7. How many fluid ounces are in 4 gallons of milk?

Solution: $$\frac{4 \text{ gallons}}{1} * \frac{4 \text{ quarts}}{1 \text{ gallon}} * \frac{2 \text{ pints}}{1 \text{ quart}} * \frac{2 \text{ cups}}{1 \text{ pint}} * \frac{8 \text{ fl. ounces}}{1 \text{ cup}}$$

$$= \frac{4}{1} * \frac{4}{1} * \frac{2}{1} * \frac{2}{1} * \frac{8 \text{ fl. ounces}}{1} = 512 \text{ fl. ounces}$$

Answer: 512 (fl. ounces)

8. If Joe needs 90 square feet of tile to complete his floor, how many square yards does he need to buy?

Solution: $$\frac{90 \text{ feet}^2}{1} * \frac{1 \text{ yard}}{3 \text{ feet}} * \frac{1 \text{ yard}}{3 \text{ feet}} = \frac{90}{1} * \frac{1 \text{ yard}}{3} * \frac{1 \text{ yard}}{3} = 10 \text{ yard}^2$$

Answer: 10 (square yards)

9. If there are 12 jegs in a jag, 4 jags in a jug, and 10 jugs in a jog, how many jegs are in a jog?

Solution: $$\frac{1 \text{ jog}}{1} * \frac{10 \text{ jugs}}{1 \text{ jog}} * \frac{4 \text{ jags}}{1 \text{ jug}} * \frac{12 \text{ jegs}}{1 \text{ jag}} = \frac{1}{1} * \frac{10}{1} * \frac{4}{1} * \frac{12 \text{ jegs}}{1} = 480 \text{ jegs}$$

Answer: 480 (jegs)

10. Pauline rides her bike at an average speed of 15 feet per second. What is her speed in miles per hour? Express your answer as a reduced improper fraction.

Solution: $\frac{15 \text{ feet}}{1 \text{ second}} * \frac{60 \text{ seconds}}{1 \text{ minute}} * \frac{60 \text{ minutes}}{1 \text{ hour}} * \frac{1 \text{ mile}}{5280 \text{ feet}} = \frac{15}{1} * \frac{60}{1} * \frac{60}{1 \text{ hour}} * \frac{1 \text{ mile}}{5280}$
$= \frac{225 \text{ mile}}{22 \text{ hour}}$

Answer: $\frac{225}{22}$ miles per hour

11. If there are 3 kones in a kylinder, 4 kylinders in a kube, and 5 kubes in a kettle, how many kones are in a kone, a kylinder, a kube and a kettle?

Solution: Convert a kylinder, a kube, and a kettle to kones then calculate the sum.

$\frac{1 \text{ kylinder}}{1} * \frac{3 \text{ kones}}{1 \text{ kylinder}} = \frac{1}{1} * \frac{3 \text{ kones}}{1} = 3 \text{ kones}$

$\frac{1 \text{ kube}}{1} * \frac{4 \text{ kylinders}}{1 \text{ kube}} * \frac{3 \text{ kones}}{1 \text{ kylinder}} = \frac{1}{1} * \frac{4}{1} * \frac{3 \text{ kones}}{1} = 12 \text{ kones}$

$\frac{1 \text{ kettle}}{1} * \frac{5 \text{ kubes}}{1 \text{ kettle}} * \frac{4 \text{ kylinders}}{1 \text{ kube}} * \frac{3 \text{ kones}}{1 \text{ kylinder}} = \frac{1}{1} * \frac{5}{1} * \frac{4}{1} * \frac{3 \text{ kones}}{1} = 60 \text{ kones}$

1 kone + 1 kylinder + 1 kube + 1 kettle = 1 kone + 3 kones + 12 kones + 60 kones = 76 kones

Answer: 76 (kones)

12. Alan's family likes to drink water. If they consume 25 gallons, 13 quarts, and 16 pints of water in a week, how many quarts of water did the family consume?

Solution: Convert 25 gallons to quarts and 16 pints to quarts then calculate the sum.

$\frac{25 \text{ gallons}}{1} * \frac{4 \text{ quarts}}{1 \text{ gallon}} = \frac{25}{1} * \frac{4 \text{ quarts}}{1} = 100 \text{ quarts}$

$\frac{16 \text{ pints}}{1} * \frac{1 \text{ quart}}{2 \text{ pints}} = \frac{16}{1} * \frac{1 \text{ quart}}{2} = 8 \text{ quarts}$

25 gallons + 13 quarts + 16 pints = 100 quarts + 13 quarts + 8 quarts = 121 quarts

Answer: 121 (quarts)

13. If a palm tree at the beach stands 2 yards, 4 feet and 5 inches tall, how tall is the tree in inches?

Solution: Convert 2 yards to inches and 4 feet to inches then calculate the sum.

$$\frac{2 \text{ yards}}{1} * \frac{3 \text{ feet}}{1 \text{ yard}} * \frac{12 \text{ inches}}{1 \text{ feet}} = \frac{2}{1} * \frac{3}{1} * \frac{12 \text{ inches}}{1} = 72 \text{ inches}$$

$$\frac{4 \text{ feet}}{1} * \frac{12 \text{ inches}}{1 \text{ feet}} = \frac{4}{1} * \frac{12 \text{ inches}}{1} = 48 \text{ inches}$$

2 yards + 4 feet + 5 inches = 72 inches + 48 inches + 5 inches = 125 inches

Answer: 125 (inches)

14. An action movie's feature-length is 2 hours, 34 minutes and 15 seconds long. How long is the movie in seconds?

Solution: Convert 2 hours to seconds and 34 minutes to seconds then calculate the sum.

$$\frac{2 \text{ hours}}{1} * \frac{60 \text{ minutes}}{1 \text{ hour}} * \frac{60 \text{ seconds}}{1 \text{ minute}} = \frac{2}{1} * \frac{60}{1} * \frac{60 \text{ seconds}}{1} = 7200 \text{ seconds}$$

$$\frac{34 \text{ minutes}}{1} * \frac{60 \text{ seconds}}{1 \text{ minute}} = \frac{34}{1} * \frac{60 \text{ seconds}}{1} = 2040 \text{ seconds}$$

2 hours + 34 minutes + 15 seconds = 7200 seconds + 2040 seconds + 15 seconds = 9255 seconds

Answer: 9255 (seconds)

15. If $1.80 = 1 Mars dollar, how much does an item priced at $8.10 cost in Mars dollars?

Solution: $$\frac{8.10 \text{ dollars}}{1} * \frac{1 \text{ Mars dollar}}{1.80 \text{ dollars}} = \frac{8.10}{1} * \frac{1 \text{ Mars dollar}}{1.80} = 4.50 \text{ Mars dollars}$$

Answer: 4.50 (Mars dollars)

16. How many cubic inches are in ⅜ of a cubic yard?

Solution: $\frac{3\text{ yard}^3}{8} * \frac{3\text{ feet}}{1\text{ yard}} * \frac{3\text{ feet}}{1\text{ yard}} * \frac{3\text{ feet}}{1\text{ yard}} * \frac{12\text{ inches}}{1\text{ feet}} * \frac{12\text{ inches}}{1\text{ feet}} * \frac{12\text{ inches}}{1\text{ feet}}$

$= \frac{3}{8} * \frac{3}{1} * \frac{3}{1} * \frac{3}{1} * \frac{12\text{ inches}}{1} * \frac{12\text{ inches}}{1} * \frac{12\text{ inches}}{1} = 17496\text{ inches}^3$

Answer: 17,496 (cubic inches)

17. The mail carrier drives his truck for 840 miles every week. If he works 8 hours per day from Monday through Friday, what is his average driving speed in miles per hour?

Solution: 8 hours for 5 days = 8 × 5 = 40 hours total

$\frac{840\text{ miles}}{40\text{ hours}} = \frac{21\text{ miles}}{1\text{ hour}}$

Answer: 21 miles per hour

18. A machine lifts 9 tons of building materials per hour. How much does it lift in ounces per second?

Solution: $\frac{9\text{ tons}}{1\text{ hour}} * \frac{2000\text{ pounds}}{1\text{ ton}} * \frac{16\text{ ounces}}{1\text{ pound}} * \frac{1\text{ hour}}{60\text{ minutes}} * \frac{1\text{ minute}}{60\text{ seconds}}$

$= \frac{9}{1} * \frac{2000}{1} * \frac{16\text{ ounces}}{1} * \frac{1}{60} * \frac{1}{60\text{ seconds}} = \frac{80\text{ ounces}}{1\text{ second}}$

Answer: 80 ounces per second

19. Anna is driving at a speed of 64 miles per hour while Kyle is driving at a speed of 49 miles per hour. How much faster is Anna driving in feet per second?

Solution: 64 miles per hour – 49 miles per hour = 15 miles per hour.

$\frac{15\text{ miles}}{1\text{ hour}} * \frac{1\text{ hour}}{60\text{ minutes}} * \frac{1\text{ minute}}{60\text{ seconds}} * \frac{5280\text{ feet}}{1\text{ mile}}$

$= \frac{15}{1} * \frac{1}{60} * \frac{1}{60\text{ seconds}} * \frac{5280\text{ feet}}{1} = \frac{22\text{ feet}}{1\text{ second}}$

Answer: 22 feet per second

20. Darry mows grass at an average rate of 660 square feet per hour. What is his average speed in square inches per minute?

Solution: $\frac{660 \text{ feet}^2}{1 \text{ hour}} * \frac{1 \text{ hour}}{60 \text{ minutes}} * \frac{12 \text{ inches}}{1 \text{ foot}} * \frac{12 \text{ inches}}{1 \text{ foot}}$

$= \frac{660}{1} * \frac{1}{60 \text{ minutes}} * \frac{12 \text{ inches}}{1} * \frac{12 \text{ inches}}{1} = \frac{1584 \text{ inches}^2}{1 \text{ minute}}$

Answer: 1,584 square inches per minute

Review Problems

1. Joey drove his car at a constant speed of 40 miles per hour for 30 minutes. How far did he travel? Express your answer in miles.

Solution: D = R × T

$$\frac{40 \text{ miles}}{1 \text{ hour}} * 30 \text{ minutes} = \frac{40 \text{ miles}}{1 \text{ hour}} * \frac{1 \text{ hour}}{60 \text{ minutes}} * 30 \text{ minutes} = 20 \text{ miles}$$

Answer: 20 miles

2. Bob drives 24 miles from his house to work. If it takes him 40 minutes to arrive at his destination, what is his average driving speed in miles per hour?

Solution: $R = \frac{D}{T}.$ $\frac{24 \text{ miles}}{40 \text{ minutes}} * \frac{60 \text{ minutes}}{1 \text{ hour}} = \frac{36 \text{ miles}}{1 \text{ hour}}$

Answer: 36 miles per hour

3. Leo works at a toy factory. If he makes toys at a constant rate of 12 toys every hour, how many toys would he make in 8 hours?

Solution: W = R × T when P = 1

$$W = \frac{12 \text{ toys}}{1 \text{ hour}} * 8 \text{ hours} = 96 \text{ toys}$$

Answer: 96 toys

4. Jarvis bikes at a constant rate and traveled 5 miles in 20 minutes. How many hours would it take him to bike 45 miles?

Solution: D = R × T; T = D ÷ R

$$T = 45 \text{ miles} \div \frac{5 \text{ miles}}{20 \text{ minutes}} = 45 \text{ miles} * \frac{20 \text{ minutes}}{5 \text{ miles}}$$

$$= 180 \text{ minutes} * \frac{1 \text{ hour}}{60 \text{ minutes}} = 3 \text{ hours}$$

Answer: 3 hours

5. Kyle sewed 525 shirts in 75 hours. What was his average rate of shirts sewed per hour?

Solution: $R = \frac{W}{T}.$ $\frac{525 \text{ shirts}}{75 \text{ hours}} = \frac{7 \text{ shirts}}{1 \text{ hour}}$

Answer: 7 shirts per hour

6. A plane travels a distance of 1125 miles at an average speed of 540 miles per hour. How many minutes does the trip take?

Solution: D = R × T; T = D ÷ R

$$T = 1125 \text{ miles} \div \frac{540 \text{ miles}}{1 \text{ hour}} = 1125 \text{ miles} * \frac{1 \text{ hour}}{540 \text{ miles}} * \frac{60 \text{ minutes}}{1 \text{ hour}} = 125 \text{ minutes}$$

Answer: 125 minutes

7. During a road trip, Jonathan drives at a rate of 45 miles per hour for the first two hours, a rate of 60 miles per hour for the next five hours and a rate of 30 miles per hour for the final three hours. What is his average speed in miles per hour for the whole road trip?

Solution: Total Distance: 45 × 2 + 60 × 5 + 30 × 3 = 90 + 300 + 90 = 480 miles
Total Time: 2 + 5 + 3 = 10 hours

$$\text{Average Speed} = \frac{480 \text{ miles}}{10 \text{ hours}} = \frac{48 \text{ miles}}{1 \text{ hour}}$$

Answer: 48 miles per hour

8. Carl can solve 14 math problems per hour. If his older brother can solve twice as many math problems per hour as Carl, how many hours will it take them to solve 336 math problems if they work together?

Solution: Carl = 14 problems per hour; Carl's brother = 14 × 2 = 28 problems per hour
Total Rate (when working together): 14 + 28 = 42 problems per hour
$T = W \div R = 336 \div 42 = 8 \text{ hours}$

Answer: 8 hours

9. Three workers can build 72 microwaves in 8 hours. How many microwaves can one worker build in one hour? Hint: It is equivalent to the rate.

Solution: W = P × R × T
72 = 3 × R × 8 = 24R. Dividing both sides by 24:
R = 72 ÷ 24 = 3 microwaves per hour

Answer: 3 microwaves per hour

10. Julian lives 120 miles from his mom. If he travels at an average speed of 40 miles per hour going to his mom's house and 24 miles per hour on the return trip, what is his average speed in miles per hour for the entire trip?

Solution: $T_1 = 120 \div 40 = 3 \text{ hours}$
$T_2 = 120 \div 24 = 5 \text{ hours}$
Total Distance: 120 + 120 = 240 miles
Total Time: 3 + 5 = 8 hours

$$\text{Average Speed} = \frac{240 \text{ miles}}{8 \text{ hours}} = \frac{30 \text{ miles}}{1 \text{ hours}}$$

Answer: 30 miles per hour

11. 6 workers can build 378 chairs in 7 hours. How many chairs can 4 workers build in 5 hours?

Solution: W = P × R × T
378 = 6 × R × 7 = 42R. Dividing both sides by 42:
R = 378 ÷ 42 = 9

W = P × R × T where P = 4, R = 9 and T = 5
W = 4 × 9 × 5 = 180 chairs

Answer: 180 chairs

12. A giant water fountain dispenses 40 quarts of water every minute. How many hours would it take to dispense 960 gallons of water? Express your answer as a mixed number.

Solution: R = $\dfrac{40 \text{ quarts}}{1 \text{ minute}} * \dfrac{1 \text{ gallon}}{4 \text{ quarts}} * \dfrac{60 \text{ minutes}}{1 \text{ hour}} = \dfrac{600 \text{ gallons}}{1 \text{ hour}}$

$$T = W \div R = 960 \text{ gallons} \div \frac{600 \text{ gallons}}{1 \text{ hour}} = 960 \text{ gallons} * \frac{1 \text{ hour}}{600 \text{ gallons}} = \frac{8}{5} \text{ hours} = 1\frac{3}{5} \text{ hours}$$

Answer: $1\frac{3}{5}$ hours

13. James has three lawns where each lawn has 1800 square feet of area. If he can mow at a constant rate of 720 square inches per minute, how many hours will it take him to finishing mowing his three lawns?

Solution: Total Area: 1800 feet^2 × 3 = 5400 feet^2

$$\frac{720 \text{ inches}^2}{1 \text{ minute}} * \frac{1 \text{ foot}}{12 \text{ inches}} * \frac{1 \text{ foot}}{12 \text{ inches}} * \frac{60 \text{ minutes}}{1 \text{ hour}} = \frac{300 \text{ feet}^2}{1 \text{ hour}}$$

$$T = W \div R = 5400 \text{ feet}^2 \div \frac{300 \text{ feet}^2}{1 \text{ hour}} = 5400 \text{ feet}^2 * \frac{1 \text{ hour}}{300 \text{ feet}^2} = 18 \text{ hours}$$

Answer: 18 hours

14. 12 workers can make 1440 DVD's in 15 hours. If each worker works at the same constant rate, 21 workers can make 840 DVD's in *X* minutes. What is *X*?

Solution:

W = P × R × T; 1440 = 12 × R × 15 = 180R. Dividing both sides by 180:
R = 1440 ÷ 180 = 8

W = P × R × T; 840 = 21 × 8 × T = 168T. Dividing both sides by 168
T = 840 ÷ 168 = 5 hours

$$5 \text{ hours} * \frac{60 \text{ minutes}}{1 \text{ hour}} = 300 \text{ minutes}$$

Answer: 300

15. Angela can feed 7 cows every hour, Bessie can feed 8 cows every hour and Cassie can feed 10 cows every hour. If they all work together, how many hours would it take them to feed 300 cows?

Solution: Working together: Total Rate = 7 + 8 + 10 = 25
T = W ÷ R = 300 ÷ 25 = 12 hours

Answer: 12 hours

16. Robin is driving ahead of Will by 20 miles traveling in the same direction. If Robin is driving at a constant speed of 39 miles per hour and Will is driving at a constant speed of 51 miles per hour, how many minutes will it take Will to catch up to Robin?

Solution: 51 mph – 39 mph = 12 mph

T = D ÷ R = $20 \text{ miles} \div \dfrac{12 \text{ miles}}{1 \text{ hour}} = 20 \text{ miles} * \dfrac{1 \text{ hour}}{12 \text{ miles}} * \dfrac{60 \text{ minutes}}{1 \text{ hour}} = 100 \text{ minutes}$

Answer: 100 minutes

17. Wally can make 6 pens in one hour. Jay can make 10 pens in half an hour. Barry can make 12 pens in 45 minutes. How many pens can they make together in 2 hours?

Solution: Wally: 6 pens in one hour × 2 = 12 pens in 2 hours
Jay: 10 pens in half an hour × 4 = 40 pens in 2 hours
Barry: $\dfrac{12 \text{ pens}}{45 \text{ minutes}} * \dfrac{60 \text{ minutes}}{1 \text{ hour}} = \dfrac{16 \text{ pens}}{1 \text{ hour}}$. 16 × 2 = 32 pens in 2 hours
Total = 12 + 40 + 32 = 84 pens

Answer: 84 pens

18. Gerald can print 2 shirts every 5 minutes. Identical twins Lucy and Lucas can each print 7 shirts every 10 minutes. How many minutes would it take for the three of them working together to print 378 shirts?

Solution: Gerald: 2 shirts every 5 minutes × 2 = 4 shirts every 10 minutes
Total Rates: 4 + 7 + 7 = 18 shirts every 10 minutes

378 ÷ 18 = 21
21 × 10 = 210 minutes

Answer: 210 minutes

19. Abby can make a button every 10 seconds, Beatrice can make a button every 15 seconds and Cassandra can make a button every 30 seconds. If they work together, how many minutes would it take them to make 408 buttons?

Solution: Abby: 1 button every 10 seconds × 3 = 3 buttons every 30 seconds
Beatrice: 1 button every 15 seconds × 2 = 2 buttons every 30 seconds
Cassandra: 1 button every 30 seconds.

Working together: 3 + 2 + 1 = 6 buttons every 30 seconds × 2 = 12 buttons every minutes.

408 ÷ 12 = 34 minutes

Answer: 34 minutes

20. Jack and Jill are 1080 miles apart. They drive at different speeds toward each other to meet up. If Jack drives at an average speed of 37 mph and Jill drives at an average speed of 53 mph, how many hours will it take them to meet up?

Solution: Combined Rates toward each other: 37 mph + 53 mph = 90 mph
T = D ÷ R = 1080 ÷ 90 = 12

Answer: 12 hours

Review Problems

1. How many two letter "words" consisting of only vowels are there?

Solution: There are 5 possibilities for each letter.
$5 \times 5 = 25$

Answer: 25 (words)

2. How many two letter "words" consisting of only consonants are there?

Solution: There are 21 possibilities for each letter.
$21 \times 21 = 441$

Answer: 441 (words)

3. How many ways can three dice of different colors be rolled?

Solution: There are 6 possibilities for each die.
$6 \times 6 \times 6 = 216$

Answer: 216 (ways)

4. How many ways can a blue die and a green die be rolled and a quarter be flipped?

Solution: There are 6 possibilities for each die and 2 possibilities for the quarter.
$6 \times 6 \times 2 = 72$

Answer: 72 (ways)

5. How many different license plates consisting of one consonant followed by two digits are there?

Solution: There are 21 possibilities for the letter and 10 possibilities for each digit.
$21 \times 10 \times 10 = 2{,}100$

Answer: 2,100 (plates)

6. How many two letter "words" consisting of a consonant followed by a vowel are there?

Solution: There are 21 possibilities for the first letter and 5 possibilities for the second letter.
$21 \times 5 = 105$

Answer: 105 (words)

7. How many five-digit numbers are there?

Solution: In order for the number to be a five-digit number, the first digit must be a nonzero digit. Thus, there are 9 choices for the first digit (1–9) and 10 choices for the last four digits (0–9).
$9 \times 10 \times 10 \times 10 \times 10 = 90{,}000$

Answer: 90,000 (numbers)

8. What is the sum of the numbers of three-digit, two-digit and one-digit numbers? (Note: 0 is a one-digit number)

Solution: This problem is simplified where the first digit can be 0 because this would include two-digit and one-digit numbers as well as three-digit numbers when the first digit is nonzero.
$10 \times 10 \times 10 = 1{,}000$

OR

Three-digit Numbers: $9 \times 10 \times 10 = 900$
Two-digit Numbers: $9 \times 10 = 90$
One-digit Numbers: 10
$900 + 90 + 10 = 1{,}000$

Answer: 1,000 (numbers)

9. Robert wants to buy a sandwich. He can choose from 3 types of bread, 5 types of meat, 4 types of vegetables and 2 types of cheese. If he picks one from each category, how many different sandwich combinations can he make?

Solution: By the Fundamental Theorem of Counting,
$3 \times 5 \times 4 \times 2 = 120$

Answer: 120 (combinations)

*10. Jackie wants to buy a sandwich. She can choose from 3 types of bread, 4 types of meat, 3 types of vegetables and 9 types of cheese. If she picks one from each category with the exception that she can omit cheese, how many different sandwich combinations can she make?

Solution: Since she can omit cheese, she has $9 + 1 = 10$ options for that category.
$3 \times 4 \times 3 \times 10 = 360$

Answer: 360 (combinations)

11. How many three letter "words" are there if the letters cannot be repeated?

Solution: For the first letter, there are 26 options. After the first letter has been chosen, there are $26 - 1 = 25$ options for the second letter. After the second letter has been chosen, there are $25 - 1 = 24$ options for the third letter.
$26 \times 25 \times 24 = 15{,}600$

Answer: 15,600 (words)

12. How many two-digit numbers are there if no digits can be repeated?

Solution: For the first digit, it must be nonzero (1–9) which means there are 9 options. The second digit has $9 - 1 = 8$ options but can also be 0 which totals to 9 options.
$9 \times 9 = 81$

Answer: 81 (numbers)

13. How many four letter "words" are there if the first letter is the same as the last letter?

Solution: For each letter except for the last letter, there are 26 options. For the last letter, there is only one option because the first letter has already been chosen from 26 options.
$26 \times 26 \times 26 \times 1 = 17{,}576$

Answer: 17,576 (words)

*14. How many three-digit numbers are there if consecutive pairs of digits are not the same digit? (i.e. 122 has a consecutive pair of "22")

Solution: The first number must be nonzero, which gives 9 options (1–9). The second digit cannot be the first digit but it can be 0 so there are 9 options. The third digit can be any digit that is not the second digit, so there are 9 options.
$9 \times 9 \times 9 = 729$

Answer: 729 (numbers)

*15. How many two letter "words" consisting of a consonant and a vowel are there? (Hint: Order matters.)

Solution: There are 21 options for the consonant and 5 options for the vowel. Since order matters, the word is either a consonant followed by a vowel or a vowel followed by a consonant. This gives 2 cases.
$2 \times 21 \times 5 = 210$

Answer: 210 (words)

16. How many five-digit even numbers can be created?

Solution: There are 9 options for the first nonzero digit(1–9). The three middle digits have 10 options (0–9). For the last digit, there are 5 options (0, 2, 4, 6, 8) that are even which will make the number even.
$9 \times 10 \times 10 \times 10 \times 5 = 45{,}000$

Answer: 45,000 (numbers)

*17. Max wants to buy a sandwich. He can choose from 5 types of bread, 7 types of meat, 6 types of vegetables and 3 types of cheese. If he is allowed to pick one from each category with the option of omitting meat, vegetable and/or cheese, how many different sandwich combinations can he make?

Solution: If he can omit meat, vegetable and/or cheese, this means there are +1 options for each of those categories.
$5 \times (7 + 1) \times (6 + 1) \times (3 + 1) = 5 \times 8 \times 7 \times 4 = 1{,}120$

Answer: 1,120 (combinations)

18. How many four-digit numbers can be created such that the number is divisible by 10 and no digit is repeated?

Solution: For the first digit, there are 9 nonzero options (1–9). The last digit must be 0 in order for the number to be divisible by 10 which gives 1 option. For the second digit, there are 8 options since two numbers are used up for the first digit and the last digit. And the third digit has 7 options because 3 digits have been used up.
$9 \times 8 \times 7 \times 1 = 504$

Answer: 504 (numbers)

19. How many license plates with three-letters followed by three-digits can be created if the first two letters are vowels, the first two digits are divisible by 3 and the last digit is divisible by 5?

Solution: The first two letters each have 5 options and the final letter has 26 options. The first two digits have 4 options (0, 3, 6, 9) that are divisible by 3 and the final digit has 2 options (0, 5) that are divisible by 5.
$5 \times 5 \times 26 \times 4 \times 4 \times 2 = 20{,}800$

Answer: 20,800 (plates)

*20. How many four letter "words" are there if consecutive pairs of letters cannot be the same letter? (i.e. book has a consecutive pair of "oo")

Solution: For the first letter, there are 26 options. For the next three letters, there are 25 options because they only need to be the letter that is not the previous letter.
$26 \times 25 \times 25 \times 25 = 406{,}250$

Answer: 406,250 (words)

Review Problems

1. How many ways can the digits in the number 1,234 be rearranged?

Solution: 4 different digits need to be arranged. There are 4! ways to arrange them.
4! = 24

Answer: 24 (ways)

2. How many ways can 6 people finish a race given that there are no ties?

Solution: 6 different people need to be arranged. There are 6! ways to arrange them.
6! = 720

Answer: 720 (ways)

3. How many ways can 8 different colored beads be placed in a circle on a table?

Solution: 8 different objects need to be arranged in a circle. There are (8 – 1)! ways to arrange them.
(8 – 1)! = 7! = 5040

Answer: 5,040 (ways)

*4. A bead bracelet consists of 7 different colored beads. How many different ways can the beads be arranged? Hint: The bracelet can be flipped.

Solution: 7 different objects need to be arranged in a circle but can also be flipped. There are $\frac{(7-1)!}{2}$ ways to arrange them.

$$\frac{(7-1)!}{2} = \frac{6!}{2} = \frac{720}{2} = 360$$

Answer: 360 (ways)

5. How many ways can 10 people enter a building if one person goes in at a time?

Solution: 10 different people need to be arranged. There are 10! ways to arrange them.
10! = 3,628,800

Answer: 3,628,800 (ways)

6. 11 people are running for office. There are three positions available: president, secretary and treasurer. One person can only hold one position. How many ways can these positions be filled up?

Solution: This is a Permutations problem because order matters (three different positions). ${}_{11}P_3 = 11 * 10 * 9 = 990$.

Answer: 990 (ways)

7. 8 people are running in a race. How many ways can the top 3 runners finish? (Assuming there are no ties)

Solution: This is a Permutations problem because order matters (three different finishes). ${}_{8}P_3 = 8 * 7 * 6 = 336$.

Answer: 336 (ways)

8. Josie has 6 favorite pens of different colors. If she can only take 3 to school, how many different sets of pens can she bring?

Solution: This is a Combinations problem because the order does not matter (the 3 pens can be chosen in any order).

$${}_{6}C_3 = \frac{6!}{3! * (6-3)!} = \frac{6 * 5 * 4 * 3!}{6 * 3!} = \frac{6 * 5 * 4}{6} = 5 * 4 = 20.$$

Answer: 20 (sets)

9. John has 10 wood blocks of different shapes. How many ways can he choose 8 blocks?

Solution: This is a Combinations problem because the order does not matter (the 8 blocks can be chosen in any order). ${}_{10}C_8 = \frac{10!}{8! * (10-8)!} = \frac{10 * 9 * 8!}{8! * 2!} = \frac{10 * 9}{2} = 45$.

Answer: 45 (ways)

10. A bag has 5 different colored marbles: red, orange, green, blue, and yellow. How many ways can 3 of the marbles be drawn from the bag if they are all drawn at the same time?

Solution: This is a Combinations problem because the order does not matter (the 3 marbles are drawn at the same time). $_5C_3 = \frac{5!}{3! * (5-3)!} = \frac{5*4*3!}{3!*2!} = \frac{5*4}{2} = 10$.

Answer: 10 (ways)

11. A company wants to choose a group of 4 representatives from 10 employees to send to a conference. How many ways can this group be formed?

Solution: This is a Combinations problem because the order does not matter (the 4 people can be chosen in any order).

$$_{10}C_4 = \frac{10!}{4! * (10-4)!} = \frac{10*9*8*7*6!}{4!*6!} = \frac{10*9*8*7}{24} = 210$$

Answer: 210 (ways)

*12. There are 4 boys and 4 girls standing in a line waiting to purchase movie tickets. How many ways can they stand in a line if all the boys must stand together and all the girls must stand together?

Solution: There are two cases to consider: the 4 boys are in the front of the line and the 4 girls are in the back of the line or the 4 girls are in the front of the line and the 4 boys are in the back of the line. For each case, there are 4! ways to arrange either the boys or the girls.

Thus, $2 * 4! * 4! = 2 * 24 * 24 = 1152$.

Answer: 1,152 (ways)

13. James has 10 favorite toys. He needs to choose 2 of them for one bag and then choose 3 of them for one box. How many ways can he choose these 5 toys to store away?

Solution: This is a Combinations problem because the order does not matter (the toys can be chosen in any order). After the first 2 toys are chosen from the 10 toys, there are 8 toys left for the last 3 to be chosen.

$$_{10}C_2 = \frac{10!}{2! * (10-2)!} = \frac{10 * 9 * 8!}{2! * 8!} = \frac{10 * 9}{2} = 45$$

$$_{8}C_3 = \frac{8!}{3! * (8-3)!} = \frac{8 * 7 * 6 * 5!}{6 * 5!} = 8 * 7 = 56$$

By the Fundamental Theorem of Counting: 45 × 56 = 2520

Answer: 2520 (ways)

14. There are 6 balls arranged in a row: 2 red, 1 green, 1 blue, 1 yellow and 1 purple. If the red balls are identical, how many ways can the balls be arranged?

Solution: $\frac{6!}{2!} = 360$

Answer: 360 (ways)

15. There are 6 balls arranged in a row: 2 red, 2 white, 1 blue and 1 yellow. If the 2 red balls are identical and the 2 white balls are identical, how many ways can the balls be arranged?

Solution: $\frac{6!}{2! * 2!} = 180$

Answer: 180 (ways)

16. How many ways can the letters in the word CREAM be arranged?

Solution: 5 different letters need to be arranged. There are 5! ways to arrange them. 5! = 120.

Answer: 120 (ways)

17. How many ways can the digits in the number 12,556 be arranged?

Solution: There are 5 digits with 5 being repeated twice. Thus, 5! is divided by 2! because the 5s can be rearranged 2! ways.

$$\frac{5!}{2!} = \frac{120}{2} = 60$$

Answer: 60 (ways)

18. How many ways can the letters in the word RACECAR be arranged?

Solution: There are 7 letters with 2 (Rs), 2 (As) and 2 (Cs).

$$\frac{7!}{2! * 2! * 2!} = \frac{5040}{8} = 630$$

Answer: 630 (ways)

19. How many ways can the digits in the number 7,844,847 be arranged?

Solution: There are 7 digits with 2 (7s), 3 (4s) and 2 (8s).

$$\frac{7!}{2! * 3! * 2!} = \frac{7 * 6 * 5 * 4 * 3!}{4 * 3!} = 210$$

Answer: 210 (ways)

20. How many ways can the letters in the word AARDVARK be arranged?

Solution: There are 8 letters with 3 (As) and 2 (Rs).

$$\frac{8!}{3! * 2!} = \frac{8 * 7 * 6 * 5 * 4 * 3!}{3! * 2} = \frac{6720}{2} = 3360$$

Answer: 3,360 (ways)

Review Problems

1. Given a set, {17, 33, 5, 25}, find the median.

Solution: Arrange the numbers: {5, 17, 25, 33}. Since there is an even number of terms, the median is the average of the two middle numbers:

$$\text{Median:} \quad = \frac{17+25}{2} = \frac{42}{2} = 21$$

Answer: 21

2. Given a set, {22, 28, 29, 22, 34}, find the mean.

Solution: There are five numbers.

$$\text{Mean:} \quad = \frac{22+28+29+22+34}{5} = \frac{135}{5} = 27$$

Answer: 27

3. Given a set, {31, 22, 58, 99, 79, 124}, find the range.

Solution: Arrange the numbers: {22, 31, 58, 79, 99, 124}.
The range is the difference between the largest and smallest value: 124 – 22 = 102

Answer: 102

4. Given a set, {17, 44, 81, 44, 99, 17, 17, 81, 99, 81}, find the mode.

Solution: Arrange the numbers: {17, 17, 17, 44, 44, 81, 81, 81, 99, 99}
17 and 81 appear 3 times while 44 and 99 appear 2 times. Thus, 17 and 81 are the modes.

Answer: 17 and 81

5. Given a set, {12, 16, 14, 14, 19, 18, 13, 14}, find the mean, median, mode and range.

Solution: Arrange the numbers: {12, 13, 14, 14, 14, 16, 18, 19}. There are 8 numbers.

$$\text{Mean:} \quad = \frac{12+13+14+14+14+16+18+19}{8} = \frac{120}{8} = 15$$

$$\text{Median:} \quad = \frac{14+14}{2} = 14$$

$$\text{Mode:} \quad = 14$$

$$\text{Range:} \quad = 19 - 12 = 7$$

Answer:

Mean: 15

Median: 14

Mode: 14

Range: 7

6. What is the average angle of a triangle with angles: 108, 44, and 28 degrees?

Solution: There are three numbers.

$$\text{Mean:} \quad = \frac{108+44+28}{3} = \frac{180}{3} = 60$$

Answer: 60 (degrees)

7. A marathon runner ran distances: 25 miles, 13 miles, 17 miles, 12 miles and 18 miles. What was the average distance that the runner ran?

Solution: There are five numbers.

$$\text{Mean:} \quad = \frac{25+13+17+12+18}{5} = \frac{85}{5} = 17$$

Answer: 17 (miles)

8. Over the course of 5 vacation days, Jake drove 23 miles, 43 miles, 37 miles, 14 miles and 13 miles. What is the median distance that he drove?

Solution: Arrange the numbers: {13, 14, 23, 37, 43}.
There is an odd number of terms, therefore, the median is the middle number.

Answer: 23 (miles)

9. Lucy works at her lemonade stand. This past week, she sold $40, $33, $18, $33, $12, $18 and $18 of lemonades. What is the positive difference between the range and mode of her daily sales?

Solution: Arrange the numbers: {12, 18, 18, 18, 33, 33, 40}
Mode = 18
Range = 40 – 12 = 28
Range – Mode = 28 – 18 = 10

Answer: 10 (dollars)

10. What is the average angle of a hexagon with angles: 120, 125, 120, 95, 136, and 124 degrees?

Solution: There are six numbers.

$$\frac{120 + 125 + 120 + 95 + 136 + 124}{6} = \frac{720}{6} = 120$$

Answer: 120 (degrees)

11. Given a set, {1, 11, 111, 1111, 11111, 111111}, find the positive difference between the median and the range.

Solution: There are six numbers.

$$\text{Median:} \ = \frac{111 + 1,111}{2} = 611$$

Range = 111,111 – 1 = 111,110
Range – Median = 111,110 – 611 = 110,499

Answer: 110,499

12. Five movies have feature-lengths of 144 minutes, 126 minutes, 125 minutes, 131 minutes and 154 minutes. What is the sum of the mean and the range of the movies' feature-lengths?

Solution: There are five numbers. Arrange the numbers: {125, 126, 131, 144, 154}

$$\text{Mean:} \ = \frac{144 + 126 + 125 + 131 + 154}{5} = \frac{680}{5} = 136$$

Range = 154 – 125 = 29
Mean + Range = 136 + 29 = 165

Answer: 165

13. Given a set, {34, 56, 71, 44, 27, 14}, find the positive difference between the mean and the range.

Solution: There are six numbers. Arrange the numbers: {14, 27, 34, 44, 56, 71}.

$$\text{Mean:} \quad = \frac{14 + 27 + 34 + 44 + 56 + 71}{6} = \frac{246}{6} = 41$$

Range = 71 – 14 = 57

Range – Mean = 57 – 41 = 16

Answer: 16

14. At the toy store, 8 different toys ranged in prices of $19, $45, $104, $99, $9, $12, $13 and $19. Let A be the mean, B be the mode and C be the median. Find the median of the set: {A, B, C}.

Solution: Arrange the numbers: {9, 12, 13, 19, 19, 45, 99, 104}

$$\text{Mean:} \quad = \frac{9 + 12 + 13 + 19 + 19 + 45 + 99 + 104}{8} = \frac{320}{8} = 40$$

Mode: 19

$$\text{Median:} \quad \frac{19 + 19}{2} = 19$$

A = 40, B = 19, C = 19

{19, 19, 40}. There are three numbers.

The median is the middle number or 19.

Answer: 19

15. Given a set, {11, A}, the mean is 8. Find the value of A.

Solution: There are 2 numbers.

(Sum of All Terms) = (Number of Terms) × (Mean)

11 + A = 2 × 8 = 16

A = 16 – 11 = 5

Answer: 5

*16. Given a set, {33, 54, 20, 12, 19, 96}, let A be the mean, B be the smallest value and C be the largest value. Find the mean of set: {A, B, C}

Solution: There are six numbers. Arrange the numbers: {12, 19, 20, 33, 54, 96}

$$\text{Mean:} \quad = \frac{12 + 19 + 20 + 33 + 54 + 96}{6} = \frac{234}{6} = 39$$

A = 39, B = 12, C = 96

{39, 12, 96}. There are three numbers.

$$\text{Mean:} \quad = \frac{39 + 12 + 96}{3} = 49$$

Answer: 49

17. Given a set, {A, B, 13, 15}, the mean is 12. Find the sum of A and B.

Solution: There are 4 numbers.
(Sum of All Terms) = (Number of Terms) × (Mean)
A + B + 13 + 15 = 4 × 12 = 48
A + B + 28 = 48
A + B = 48 – 28 = 20

Answer: 20

18. Three trees have heights of A, B, and 10 feet. If the mode of the trees' heights is 7 feet, what is the mean of the trees' heights?

Solution: Since the mode of the heights is 7 feet and there are 3 terms where one of them is 10, A and B must both equal 7 because there must be a repeat term for there to be a mode. A = B = 7.

{7, 7, 10}. There are three numbers.

$$\text{Mean:} \quad = \frac{7 + 7 + 10}{3} = \frac{24}{3} = 8$$

Answer: 8 (feet)

19. Given an arranged set, {14, 16, A, 20, 22, B}, the median and the mean is 20. Find the positive difference between A and B.

Solution: There are six numbers. Since the set is arranged in order, A and 20 are the two middle numbers. Thus, A can be calculated using the median.

$\frac{A+20}{2} = 20 \qquad A + 20 = 20 * 2 = 40 \qquad A = 40 - 20 = 20$

Now that A has been found, B can be found using the mean.

(Sum of All Terms) = (Number of Terms) × (Mean)
(Sum of All Terms) = 14 + 16 + A + 20 + 22 + B = 72 + A + B = 72 + 20 + B = 92 + B
(Number of Terms) × (Mean) = 6 × 20 = 120
92 + B = 120
B = 120 – 92 = 28

B – A = 28 – 20 = 8

Answer: 8

*20. Given an arranged set, {10, 16, 17, A, B, 23, C}, the mean and median is 19 and the range is 18. Find the product of the median and range of set: {A, B, C}

Solution: There are 7 numbers.
Thus, the middle number is the median, which in this case is A = 19.
The range is C - 10 = 18, in which C = 18 + 10 = 28.
Now, the mean can be used to find B.
(Sum of All Terms) = (Number of Terms) × (Mean)
(Sum of All Terms) = 10 + 16 + 17 + A + B + 23 + C = 10 + 16 + 17 + 19 + B + 23 + 28
= 113 + B
(Number of Terms) × (Mean) = 7 × 19 = 133
113 + B = 133
B = 133 – 113 = 20

{19, 20, 28}. There are three numbers.
Median = 20
Range = 28 – 19 = 9
20 × 9 = 180.

Answer: 180

Review Problems

1. If a die is rolled, what is the probability the number rolled is even?

Solution: 3 outcomes are even (2, 4 and 6)

$$\text{P(even)} = \frac{3}{6} = \frac{1}{2}$$

Answer: ½

2. If two coins are flipped, what is the probability that there is no tail?

Solution: $$\text{P(0 tails)} = \text{P(no tail)} * \text{P(no tail)} = \frac{1}{2} * \frac{1}{2} = \frac{1}{4}$$

Answer: ¼

3. A bag contains 5 blue, 11 green, 4 yellow and 2 orange marbles. If a marble is chosen at random, what is the probability that it is not a green marble?

Solution: $$\text{P(Green)} = \frac{11}{5 + 11 + 4 + 2} = \frac{11}{22} = \frac{1}{2}$$

$$\text{P(Not Green)} = 1 - \text{P(Green)} = 1 - \frac{1}{2} = \frac{1}{2}$$

Answer: ½

4. A die is rolled 4 times. The results are 3, 3, 6 and 4. If the die is rolled again, what is the probability that it rolls a 2?

Solution: The probability is still ⅙ regardless of the results from previous rolls.

Answer: ⅙

5. If a die is rolled and a coin is flipped, what is the probability the coin flips a head and the die rolls a 4?

Solution: $$P = \frac{1}{2} * \frac{1}{6} = \frac{1}{12}$$

Answer: $\frac{1}{12}$

6. If two dice are rolled, what is the probability that the sum of the numbers rolled is 1?

Solution: The minimum number rolled for a die is 1 so the minimum sum for two dice rolled is 1 + 1 = 2. Thus, rolling a sum of 1 is impossible. P(Sum = 1) = 0

Answer: 0

7. If a number is randomly chosen from 1 to 10, inclusive, what is the probability the number chosen is a prime number?

Solution: There are 4 prime numbers between 1 and 10, inclusive: 2, 3, 5 and 7

$$\text{P(Prime Number)} = \frac{4}{10} = \frac{2}{5}$$

Answer: ⅖

8. If two dice are rolled, what is the probability that the sum of the numbers rolled is 2?

Solution: There is only 1 case where the sum is 2: when both dice are 1s.
And there are 6 × 6 = 36 possible outcomes.

$$\text{P(Sum} = 2) = \frac{1}{36}$$

Answer: $\frac{1}{36}$

9. A bag has 20 marbles in which ½ are blue, ⅕ are yellow and the rest are red. If 7 blue marbles, 5 yellow marbles and 6 red marbles are added to the bag, what is the probability that a blue marble is not drawn?

Solution: $\text{Blue} = 20 * \frac{1}{2} = 10$. $\text{Yellow} = 20 * \frac{1}{5} = 4$. $\text{Red} = 20 - 10 - 4 = 6.$

New Blue = 10 + 7 = 17; New Yellow = 4 + 5 = 9; New Red = 6 + 6 = 12

$$\text{P(Not Blue)} = \frac{\text{Yellow} + \text{Red}}{\text{Total}} = \frac{9+12}{17+9+12} = \frac{21}{38}$$

Answer: $\frac{21}{38}$

10. If two coins are flipped, what is the probability that there is at least one head?

Solution: $\text{P(0 heads)} = \text{P(no head)} * \text{P(no head)} = \frac{1}{2} * \frac{1}{2} = \frac{1}{4}$

$$\text{P(At least 1 head)} = 1 - \text{P(0 heads)} = 1 - \frac{1}{4} = \frac{3}{4}$$

Answer: ¾

11. What is the probability that a number chosen from 1 to 100, inclusive, is not divisible by 14?

Solution: 100 ÷ 14 = 7 R 2. There are 7 multiples of 14 from 1 to 100.

$$P = \frac{100-7}{100} = \frac{93}{100}$$

Answer: $\frac{93}{100}$

12. A bookshelf holds 14 math books, 13 history books and 11 science books. If 2 books from each subject are removed from the shelf, what is the probability that a book chosen randomly from the shelf is a history book?

Solution:

New Math = 14 – 2 = 12; New History = 13 – 2 = 11; New Science = 11 – 2 = 9

$$\text{P(history)} = \frac{\text{History}}{\text{Total}} = \frac{11}{12+11+9} = \frac{11}{32}$$

Answer: $\frac{11}{32}$

13. Jake has a bag of 29 gumballs. There are 5 red gumballs, three times as much green gumballs as red gumballs and the remaining gumballs are blue. If he randomly chooses one gumball from the bag, what is the probability that the gumballs is blue?

Solution: Red = 5; Green = 5 × 3 = 15; Blue = 29 – 5 – 15 = 9

$$\text{P(blue)} = \frac{\text{Blue}}{\text{Total}} = \frac{9}{29}$$

Answer: $\frac{9}{29}$

14. If three coins are flipped, what is the probability that there are two heads and one tail?

Solution: There are 2 × 2 × 2 = 8 possible outcomes.

The number of ways the 2 heads and 1 tail can be arranged is: $\frac{3!}{2! * 1!} = 3$

$$\text{P(2 heads and 1 tail)} = \frac{3}{8}$$

Answer: ⅜

15. A coin is flipped and two dice are rolled. What is the probability the coin is not heads and the sum of the numbers rolled on the dice are 12?

Solution: There is only 1 case where the sum is 12: when both dice are 6s
And there are 6 × 6 = 36 possible outcomes for the two dice.

$$P = \text{P(not head)} * \text{P(Sum} = 12) = \frac{1}{2} * \frac{1}{36} = \frac{1}{72}$$

Answer: $\frac{1}{72}$

*16. A bag has 10 marbles: 5 red, 3 yellow and 2 blue. What is the probability that a red marble is drawn first then a yellow marble is drawn second without replacement?

Solution: $$P = \text{P(Red)} * \text{P(Yellow)} = \frac{5}{10} * \frac{3}{10-1} = \frac{1}{2} * \frac{3}{9} = \frac{1}{6}$$

Answer: ⅙

17. If four coins are flipped, what is the probability that there is at least one head?

Solution: $\text{P(0 heads)} = \text{P(no head)}^4 = (\frac{1}{2})^4 = \frac{1}{16}$

$\text{P(At least 1 head)} = 1 - \text{P(0 heads)} = 1 - \frac{1}{16} = \frac{15}{16}$

Answer: $\frac{15}{16}$

18. What is the probability that a number chosen from 1 to 100, inclusive, is divisible by 3 but not divisible by 9?

Solution: 100 ÷ 3 = 33 R 1; 100 ÷ 9 = 11 R 1.
33 – 11 = 22

$P = \frac{22}{100} = \frac{11}{50}$

Answer: $\frac{11}{50}$

*19. If three dice are rolled, what is the probability that either all 3 numbers are odd or all 3 numbers are even?

Solution: $\text{P(3 odd)} = \text{P(odd)}^3 = (\frac{1}{2})^3 = \frac{1}{8}$

$\text{P(3 even)} = \text{P(even)}^3 = (\frac{1}{2})^3 = \frac{1}{8}$

$P = \frac{1}{8} + \frac{1}{8} = \frac{1}{4}$

Answer: ¼

*20. What is the probability that a number chosen from 1 to 1000, inclusive, is divisible by 10 but not divisible by 3?

Solution: 1000 ÷ 10 = 100 R 0
10 × 3 = 30
1000 ÷ 30 = 33 R 10

100 – 33 = 67

$$P = \frac{67}{1000}$$

Answer: $\frac{67}{1000}$

Review Problems

1. What are the three angles of an equilateral triangle?

Solution: An equilateral triangle has all equal sides and all equal angles. If there are three equal angles that total to 180 degrees, each one must be $180 \div 3 = 60$ degrees or 60°.

Answer: 60°, 60°, 60°

2. Find the perimeter of an equilateral triangle with side length 7.

Solution: An equilateral triangle has all equal sides and the perimeter is the total length around a figure. Since there are 3 equal sides of length 7, the perimeter is $p = 3s = 7 \times 3 = 21$.

Answer: 21

3. One of a right triangle's angles is 32°. What are the measures of its other two angles?

Solution: A right triangle always has a 90° angle. If one of the angles in a right triangle is 32°, the other angle is $180° - 90° - 32° = 58°$.

Faster Way: Non-right angles in a right triangle are complementary. Thus, $90° - 32° = 58°$ yields the other angle.

Answer: 58°, 90°

4. Find the side length of an equilateral triangle with perimeter 93.

Solution: Since an equilateral triangle has 3 equal sides, the side length is $s = p \div 3 = 93 \div 3 = 31$.

Answer: 31

5. Find the perimeter of a scalene triangle with side lengths: 3, 5 and 7.

Solution: The perimeter is the sum of all the side lengths: $p = a + b + c = 3 + 5 + 7 = 15$.

Answer: 15

6. One of the base angles of an isosceles triangle is $64°$. What are the other two angles?

Solution: In an isosceles triangle, there are two base angles. Thus, one of the other two angles is $64°$. Since the sum of all three angles is $180°$, the third angle can be found: $180° - 64° - 64° = 52°$.

Answer: 52°, 64°

7. Find the perimeter of a square with side length 13.

Solution: A square has four equal side lengths. The perimeter is $p = 4s = 4 \times 13 = 52$.

Answer: 52

8. What are the measurements of all four angles of a square?

Solution: A square has four equal angles. If the angles total to $360°$, each of them is $360° \div 4 = 90°$.

Answer: 90°, 90°, 90°, 90°

9. Find all of the side lengths of a square with perimeter 104.

Solution: A square has four equal side lengths. Each of the side lengths is, $s = p \div 4 = 104 \div 4 = 26$.

Answer: 26, 26, 26, 26

10. A square has the same perimeter as an equilateral triangle. If the side length of the triangle is 24, what is the side length of the square?

Solution: The perimeter of the equilateral triangle is $p = 3s = 3 \times 24 = 72$. Since the square and the triangle have the same perimeter, the perimeter of the square is also 72. $p = 4s$. $s = p \div 4 = 72 \div 4 = 18$.

Answer: 18

11. Two equal angles are supplementary. Find one of the two angles.

Solution: If two equal angles add up to 180°, each of them must be $180° \div 2 = 90°$.

Answer: 90°

12. A quadrilateral has a perimeter 80 and three of its sides have lengths 21, 25 and 18. What is the fourth side length?

Solution: The fourth side length can be found by subtracting the 3 other sides from the perimeter.
$80 - 21 - 25 - 18 = 16$

Answer: 16

13. Find all of the side lengths of a rhombus with perimeter 56.

Solution: A rhombus has four equal sides. $p = 4s$. $s = p \div 4 = 56 \div 4 = 14$.

Answer: 14, 14, 14, 14

14. Two equal angles are complementary. Find one of the two angles.

Solution: If two equal angles add up to 90°, each of them must be $90° \div 2 = 45°$.

Answer: 45°

15. If one of a rhombus' angles is 23°, find the other three angles.

Solution: Since opposite angles of a rhombus are equal, one of the angles is also 23°. Adjacent angles are supplementary so one of the other two angles is $180° - 23° = 157°$. And since, opposite angles are equal, the last angle must be 157°.

Answer: 23°, 157°, 157°

16. Find the length of a rectangle if the perimeter is 84 and the width is 13.

Solution: Perimeter: $p = 2l + 2w$. $(p = 84, w = 13)$
$84 = 2l + 2 \times 13 = 2l + 26$
$2l = 58$ $l = 58 \div 2 = 29$

Answer: 29

17. A parallelogram, a square and an equilateral triangle have the same perimeter. If the 2 different side lengths of the parallelogram are 84 and 120, what is the sum of the side length of the square and the side length of the equilateral triangle?

Solution: The perimeter of the parallelogram is $p = 2a + 2b = 2 \times 84 + 2 \times 120 = 408$.
If the parallelogram has the same perimeter as the square and the equilateral triangle, then the side length of the square is $408 \div 4 = 102$ and the side length of the triangle is $408 \div 3 = 136$. $102 + 136 = 238$.

Answer: 238

*18. The vertex angle of an isosceles triangle is 24°. If the base angle of the isosceles triangle is the same as one of the angles in a rhombus, find all of the angles of the rhombus.

Solution: If the vertex angle is 24°, the sum of the two equal base angles is $180° - 24° = 156°$. This means that one of the base angles is $156° \div 2 = 78°$.

If this is one of the angles in the rhombus, then two of the angles are 78°. And since adjacent/different angles are supplementary, the other two angles are $180° - 78° = 102°$.

Answer: 78°, 78°, 102°, 102°

*19. What is the supplement of the complement of the supplement of 148°?

Solution: First, work backwards and find the supplement of 148°. $180° - 148° = 32°$.
Now, the question is simplified to "the supplement of the complement of 32°".
Second, find the complement of 32°. $90° - 32° = 58°$.
Now, the question is simplified to "the supplement of 58°".
Lastly, find the supplement of 58°. $180° - 58° = 122°$

Answer: 122°

20. A rhombus and a rectangle have the same perimeter. The side lengths of the rectangle are 13 and 85. What are the side lengths of the rhombus?

Solution: The perimeter of the rectangle is $p = 2l + 2w = 2 \times 13 + 2 \times 85 = 196$. Since a rhombus has 4 equal side lengths, each of its side lengths is $196 \div 4 = 49$.

Answer: 49, 49, 49, 49

Review Problems

1. What is the area of a square with side length 10?

Solution: $A = s^2 = 10^2 = 10 \times 10 = 100$

Answer: 100

2. What is the length of a rectangle with area 80 and width 10?

Solution: $A = l \times w$; $80 = l \times 10$
$l = 80 \div 10 = 8$

Answer: 8

3. What is the area of a triangle with base 16 and height 20? (Note: The height is perpendicular to the base)

Solution: $A = \frac{1}{2}bh = \frac{1}{2} * 16 * 20 = 160$

Answer: 160

4. What is the circumference of a circle with diameter 16? Express your answer in terms of π.

Solution: $C = \pi d = \pi * 16 = 16\pi$

Answer: 16π

5. What is the area of a circle with diameter 22? Express your answer in terms of π.

Solution: $d = 22$; $d = 2r$
$r = 22 \div 2 = 11$
$A = \pi r^2 = \pi * 11^2 = 121\pi$

Answer: 121π

6. A triangle and a square have the same area. If the base of the triangle is 12 and the side length of the square is 6, what is the height of the triangle? (Note: The height is perpendicular to the base)

Solution: $A_{\text{square}} = s^2 = 6^2 = 36$

$A_{\text{triangle}} = \frac{1}{2}bh = \frac{1}{2} * 12 * h = 6h = 36$

h = 36 ÷ 6 = 6

Answer: 6

7. A square and a rectangle have the same area. If the rectangle's length is 9 and its width is 4, what is the side length of the square?

Solution: $A_{\text{rectangle}} = lw = 9 * 4 = 36$

$A_{\text{square}} = s^2 = 36$

$s = \sqrt{36} = 6$

Answer: 6

*8. An isosceles right triangle has the same area as a rectangle with side lengths 10 and 5. What is the length of one of the legs of the right triangle?

Solution: An isosceles right triangle has two equal legs.

$A_{\text{rectangle}} = lw = 10 * 5 = 50$

$A_{\text{triangle}} = \frac{1}{2}l^2 = 50$. Multiply 2 to both sides.

$l^2 = 2 \times 50 = 100$

$l = \sqrt{100} = 10$

Answer: 10

9. Find the area of a rectangle with length 10 and perimeter 30.

Solution: $p = 2l + 2w = 2 * 10 + 2w = 2w + 20 = 30$. Subtract 20 from both sides:

2w = 10 w = 10 ÷ 2 = 5

A = l × w = 10 × 5 = 50

Answer: 50

10. What is the area of a right triangle with legs 11 and 60, and hypotenuse 61?

Solution: $A_{\text{triangle}} = \frac{1}{2}l_1l_2 = \frac{1}{2} * 11 * 60 = 330$

Answer: 330

11. Find the circumference of a circle with area 81π. Express your answer in terms of π.

Solution: $A = \pi r^2 = 81\pi$. Divide both sides by π:
$r^2 = 81$
$r = \sqrt{81} = 9$
$C = 2\pi r = 2 * \pi * 9 = 18\pi$

Answer: 18π

12. Given a rectangle with area 91 where the side lengths are positive integers greater than 1, find its perimeter.

Solution: Prime Factorization of 91: $91 = 7^1 13^1$.
l = 13 and w = 7
$p = 2l + 2w = 2 * 13 + 2 * 7 = 40$

Answer: 40

13. A square's side length has the same length as the radius of a circle with circumference 24π. What is the area of the square?

Solution: $C = 2\pi r = 24\pi$. Divide both sides by 2π:
r = 24π ÷ 2π = 12
r = s = 12

$A_{\text{square}} = s^2 = 12^2 = 144$

Answer: 144

14. What is the positive difference between the numerical value of circumference and the area of a circle with diameter 18? Express your answer in terms of π.

Solution: d = 18; d = 2r

r = 18 ÷ 2 = 9

$C = 2\pi r = 2 * \pi * 9 = 18\pi$

$A = \pi r^2 = \pi * 9^2 = 81\pi$

A – C = $81\pi - 18\pi = 63\pi$

Answer: 63π

15. Find the area of the circle where its radius is equivalent to the width of rectangle with perimeter 26 and length 9. Express your answer in terms of π.

Solution: $p = 2l + 2w = 2 * 9 + 2w = 2w + 18 = 26$. Subtract both sides by 18:

2w = 26 – 18 = 8

w = 8 ÷ 2 = 4.

r = w = 4

$A = \pi r^2 = \pi * 4^2 = 16\pi$

Answer: 16π

16. A rectangle's area is numerically equivalent to a square's perimeter. If the side lengths of the rectangle are 10 and 14, what is the side length of the square?

Solution: $A_{\text{rectangle}} = l * w = 10 * 14 = 140$

$P_{\text{square}} = 4s = 140$

s = 140 ÷ 4 = 35

Answer: 35

17. A right triangle has area 180. If one of the legs is 40, find the other leg.

Solution: $A_{\text{triangle}} = \frac{1}{2}l_1 l_2 = \frac{1}{2} * 40 * l_2 = 20l_2 = 180$. Divide both sides by 20:

$l_2 = 180 \div 20 = 9$

Answer: 9

18. A large square consists of 16 smaller squares (4 by 4) each with side length 5. Find the area of the large square.

Solution: $A_{\text{small square}} = 5 * 5 = 25$

$A_{\text{large square}} = 25 * 16 = 400$

Answer: 400

*19. The first circle has diameter 16. The second circle has a circumference equivalent to the sum of the numerical value of the first circle's circumference and area. Find the area of the second circle.

Solution: $d = 2r$ $\quad$ $r = d \div 2$ $\quad$ $r_1 = 16 \div 2 = 8$

$C_2 = A_1 + C_1 = \pi * 8^2 + 2 * \pi * 8 = 64\pi + 16\pi = 80\pi$

$C_2 = 2\pi * r_2 = 80\pi$. Divide both sides by 2π:

$r_2 = 80\pi \div 2\pi = 40$

$A_2 = \pi r^2 = \pi * 40^2 = 1600\pi$

Answer: 1600π

*20. An isosceles triangle has the same area as a rectangle with length 12 and width 5. If the legs of the triangle are 13 and the height is 12, what is the perimeter of the triangle?

Solution: $A_{rectangle} = 12 * 5 = 60$

$A_{triangle} = \frac{1}{2}bh = \frac{1}{2} * b * 12 = 6b = 60$. Divide both sides by 6:

$b = 60 \div 6 = 10$

$p = b + l + l = 10 + 13 + 13 = 36$

Answer: 36

Review Problems

1. What is the perimeter of an equilateral quadrilateral with side length 5?

Solution: $p = n \times s = 5 \times 4 = 20$

Answer: 20

2. A triangle has perimeter 25 with side lengths 12 and 5. What is the third side length of the triangle?

Solution: $25 - 12 - 5 = 8$

Answer: 8

3. Find one of the angles of an equiangular hexagon.

Solution: $$180 - \frac{360}{n} = 180 - \frac{360}{6} = 180 - 60 = 120$$

Answer: 120 degrees

4. If three of the side lengths of a hexagon are 1 and the other three side lengths are 2, what is the perimeter of the hexagon?

Solution: $1 + 1 + 1 + 2 + 2 + 2 = 9$

Answer: 9

5. Find one of the angles of a regular decagon.

Solution: $$180 - \frac{360}{n} = 180 - \frac{360}{10} = 180 - 36 = 144$$

Answer: 144 degrees

6. Find the side length of an regular octagon with perimeter 24.

Solution: p = n × s
s = p ÷ n = 24 ÷ 8 = 3

Answer: 3

7. A hexagon has two different angle measures. Three of its angles are 100 degrees and the other three angles are *X* degrees. What is the value of *X*?

Solution: Sum = $(n-2)*180=(6-2)*180=4*180=720$

$$\frac{(720-3*100)}{3}=\frac{420}{3}=140$$

Answer: 140

8. What is the sum of all the angles of a pentagon?

Solution: Sum = $(n-2)*180=(5-2)*180=3*180=540$

Answer: 540 degrees

9. How many diagonals does a pentagon have?

Solution: d = $\frac{n*(n-3)}{2}=\frac{5*(5-3)}{2}=\frac{5*2}{2}=5$

Answer: 5 diagonals

10. Find the mean of all of the angles in a nonagon.

Solution: Sum = $(n-2)*180=(9-2)*180=7*180=1260$

$$\frac{1260}{9}=140$$

Answer: 140 degrees

11. Find the positive difference between the side length of a regular pentagon with perimeter 35 and the side length of a regular heptagon with perimeter 21.

Solution: Pentagon: s = p ÷ 5 = 35 ÷ 5 = 7
Heptagon: s = p ÷ 7 = 21 ÷ 7 = 3

7 – 3 = 4

Answer: 4

12. A decagon has side lengths: 1, 2, 3, 4, 5, 6, 7, 8, 9 and 10. If an equilateral pentagon has the same perimeter as the decagon, what is the side length of the pentagon?

Solution: Decagon: P = 1 + 2 + 3 + 4 + 5 + 6 + 7 + 8 + 9 + 10 = 55
Pentagon: s = p ÷ n = 55 ÷ 5 = 11

Answer: 11

13. How many diagonals does a heptagon have?

Solution: d = $\frac{n * (n-3)}{2} = \frac{7 * (7-3)}{2} = \frac{7 * 4}{2} = 14$

Answer: 14 diagonals

14. How many more diagonals does a 19-gon have than a 18-gon?

Solution: 19-gon: d = $\frac{n * (n-3)}{2} = \frac{19 * (19-3)}{2} = \frac{19 * 16}{2} = 152$

18-gon: d = $\frac{n * (n-3)}{2} = \frac{18 * (18-3)}{2} = \frac{18 * 15}{2} = 135$

152 – 135 = 17

Answer: 17 diagonals

15. How many more diagonals does a 20-gon have than a 19-gon?

Solution: 20-gon: d = $\frac{n*(n-3)}{2} = \frac{20*(20-3)}{2} = \frac{20*17}{2} = 170$

19-gon: d = $\frac{n*(n-3)}{2} = \frac{19*(19-3)}{2} = \frac{19*16}{2} = 152$

170 – 152 = 18

Answer: 18 diagonals

16. A regular quadrilateral, a regular pentagon, a regular hexagon and a regular heptagon have the same perimeter. If the side length of the hexagon is 105, what is the sum of the side lengths of the quadrilateral, pentagon, hexagon and heptagon?

Solution: Hexagon: p = 6 × s = 6 × 105 = 630

$s_{quadrilateral} = 630 \div 4 = \frac{315}{2}$ = 157.5

$s_{pentagon} = 630 \div 5 = 126$

$s_{heptagon} = 630 \div 7 = 90$

$\frac{315}{2} + 105 + 126 + 90 = \frac{957}{2}$ = 478.5

Answer: $\frac{957}{2}$ or 478.5

17. Find the positive difference between one of the angles in a regular 20-gon and one of the angles in a regular 30-gon.

Solution: 20-gon: $180 - \frac{360}{n} = 180 - \frac{360}{20} = 180 - 18 = 162$

30-gon: $180 - \frac{360}{n} = 180 - \frac{360}{30} = 180 - 12 = 168$

168 – 162 = 6

Answer: 6 degrees

18. The numerical value of one of the angles in a regular octagon is equivalent to one of its side lengths. What is the perimeter of this octagon?

Solution: The sum of the eight angles will be equal to the perimeter or the sum of the 8 sides.
Sum = $(n-2)*180 = (8-2)*180 = 6*180 = 1080$

Answer: 1080

19. Find the sum of one of the angles in a regular 360-gon and one of the angles in a regular 180-gon.

Solution: 360-gon: $180 - \frac{360}{n} = 180 - \frac{360}{360} = 180 - 1 = 179$

180-gon: $180 - \frac{360}{n} = 180 - \frac{360}{180} = 180 - 2 = 178$

179 + 178 = 357

Answer: 357 degrees

20. A hexagon's angles form an arithmetic sequence. If the smallest angle is 95 degrees and the largest angle is 145 degrees, find the other four angles.

Solution: $a_1 = 95, a_6 = 145$. These terms are separated by 5 common differences (6 – 1 = 5).
5d = 145 – 95 = 50. Divide both sides by 5:
d = 50 ÷ 5 = 10

$a_2 = a_1 + d = 95 + 10 = 105$
$a_3 = a_2 + d = 105 + 10 = 115$
$a_4 = a_3 + d = 115 + 10 = 125$
$a_5 = a_4 + d = 125 + 10 = 135$

Answer: 105, 115, 125, and 135 degrees

Review Problems

1. Find the hypotenuse of a right triangle with legs 6 and 8.

Solution: $c^2 = a^2 + b^2 = 6^2 + 8^2 = 36 + 64 = 100$
$c = \sqrt{100} = 10$

Answer: 10

2. Find the hypotenuse of a right triangle with legs 300 and 400.

Solution: $c^2 = 300^2 + 400^2 = 90000 + 160000 = 250000$
$c = \sqrt{250000} = 500$

Answer: 500

3. Find the hypotenuse of an isosceles right triangle with leg s. Express your answer in terms of s. Hint: An isosceles right triangle has 2 equal legs.

Solution: $c^2 = s^2 + s^2 = 2s^2$
$c = \sqrt{2s^2} = s\sqrt{2}$

Answer: $s\sqrt{2}$ (Note: An isosceles right triangle's sides has ratio $s : s : s\sqrt{2}$)

4. Find the hypotenuse of an isosceles right triangle with legs 10 and 10.

Solution: Using the answer from problem #3, $s = 10$
$c = s\sqrt{2} = 10\sqrt{2}$

Answer: $10\sqrt{2}$

5. Find the hypotenuse of a right triangle with legs 4 and 6.

Solution: $c^2 = 4^2 + 6^2 = 16 + 36 = 52$
$c = \sqrt{52} = \sqrt{2^2 13^1} = 2\sqrt{13}$

Answer: $2\sqrt{13}$

6. Find the other leg of a right triangle with leg $\sqrt{15}$ and hypotenuse $\sqrt{31}$.

Solution: $a^2 = c^2 - b^2 = (\sqrt{31})^2 - (\sqrt{15})^2 = 31 - 15 = 16$
$a = \sqrt{16} = 4$

Answer: 4

7. Find the perimeter of a right triangle with legs $4\sqrt{2}$ and $2\sqrt{2}$.

Solution: $c^2 = (4\sqrt{2})^2 + (2\sqrt{2})^2 = 32 + 8 = 40$
$c = \sqrt{40} = 2\sqrt{10}$
$p = a + b + c = 4\sqrt{2} + 2\sqrt{2} + 2\sqrt{10} = 6\sqrt{2} + 2\sqrt{10}$

Answer: $6\sqrt{2} + 2\sqrt{10}$

8. Find the area of a rectangle with a length $2\sqrt{55}$ and a diagonal 20.

Solution: $w^2 = 20^2 - (2\sqrt{55})^2 = 400 - 220 = 180$
$w = \sqrt{180} = 6\sqrt{5}$
$A = lw = 2\sqrt{55} * 6\sqrt{5} = 60\sqrt{11}$

Answer: $60\sqrt{11}$

9. Find the other leg of a right triangle with leg 84 and hypotenuse 85.

Solution: $a^2 = c^2 - b^2 = (c - b) * (c + b) = 85^2 - 84^2 = (85 - 84) * (85 + 84)$
$= 1 * 169 = 169$
$a = \sqrt{169} = 13$

Answer: 13

10. Find the area of a right triangle with leg 48 and hypotenuse 50.

Solution: $a^2 = c^2 - b^2 = (c - b) * (c + b) = (50 - 48) * (50 + 48) = 2 * 98 = 196$
$a = \sqrt{196} = 14$

$$A = \frac{1}{2}l_1 l_2 = \frac{1}{2} * 14 * 48 = 336$$

Answer: 336

11. Find the perimeter of a square with a diagonal $\sqrt{14}$.

Solution: $s^2 + s^2 = 2s^2 = (\sqrt{14})^2 = 14$

$s^2 = 7;$ $\quad s = \sqrt{7}$

$p = 4s = 4\sqrt{7}$

Answer: $4\sqrt{7}$

12. Find the area of a square with a diagonal *d*. Express your answer in terms of *d*.

Solution: $s^2 + s^2 = 2s^2 = d^2$

$$s^2 = \frac{d^2}{2}$$

Since Area = A = s^2,

$$A = s^2 = \frac{d^2}{2}$$

Answer: $\frac{d^2}{2}$

13. Find the area of a square with a diagonal $\sqrt{78}$.

Solution: Using the answer from problem #12, $A = \frac{d^2}{2} = \frac{(\sqrt{78})^2}{2} = \frac{78}{2} = 39$

Answer: 39

14. Find the perimeter of the largest right triangle in the following figure. (Figure not drawn to scale)

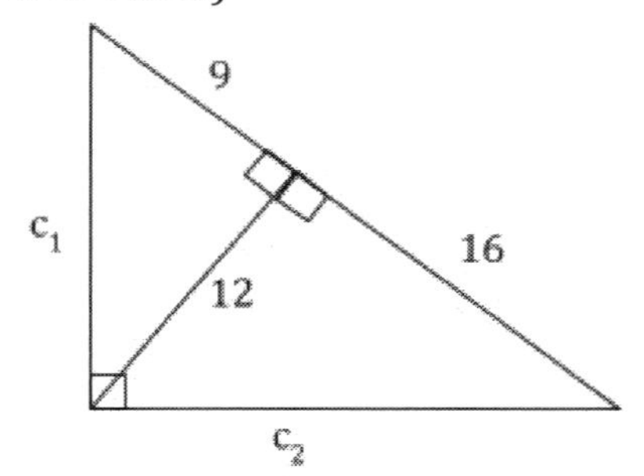

Solution: The top right triangle has legs 9 and 12: $c_1^2 = 9^2 + 12^2 = 225$
$c_1 = \sqrt{225} = 15$

The bottom right triangle has legs 12 and 16: $c_2^2 = 12^2 + 16^2 = 400$
$c_2 = \sqrt{400} = 20$

Thus, the largest right triangle has legs 15 and 20 and hypotenuse 9 + 16 = 25.
P = 15 + 20 + 25 = 60

Answer: 60

15. Find the perimeter of a right triangle with leg 11 and hypotenuse 61.

Solution: $a^2 = 61^2 - 11^2 = (61 - 11) * (61 + 11) = 50 * 72$
$a = \sqrt{50 * 72} = 60$
P = 11 + 60 + 61 = 132

Answer: 132

16. Find the area of the following quadrilateral (Figure not drawn to scale):

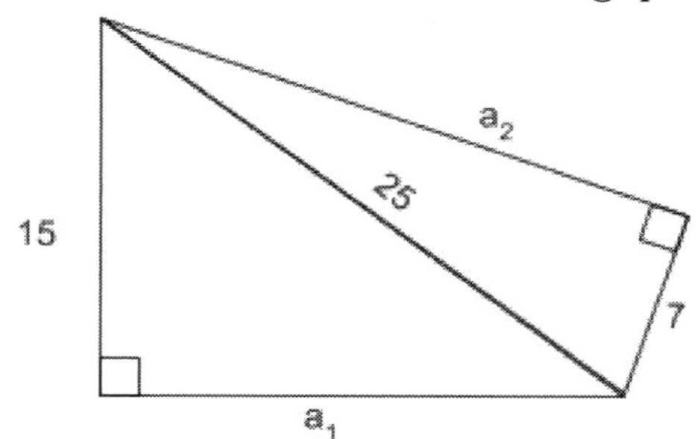

Solution: The right triangle on the left has leg 15 and hypotenuse 25:
$a_1^2 = 25^2 - 15^2 = (25 - 15) * (25 + 15) = 10 * 40 = 400$

$a_1 = \sqrt{400} = 20$

The right triangle on the right has leg 7 and hypotenuse 25:
$a_2^2 = 25^2 - 7^2 = (25 - 7) * (25 + 7) = 18 * 32$

$a_2 = \sqrt{18 * 32} = 24$

Area of the first triangle: $A = \frac{1}{2} l_1 l_2 = \frac{1}{2} * 15 * 20 = 150$

Area of the second triangle: $A = \frac{1}{2} l_1 l_2 = \frac{1}{2} * 7 * 24 = 84$

Area of the quadrilateral: 150 + 84 = 234

Answer: 234

17. Find the perimeter of an isosceles right triangle with a leg *s*. Express your answer in terms of *s*. Hint: An isosceles right triangle has 2 equal legs.

Solution: $c^2 = s^2 + s^2 = 2s^2$

$c = \sqrt{2s^2} = s\sqrt{2}$

$p = s + s + s\sqrt{2} = 2s + s\sqrt{2}$

Answer: $2s + s\sqrt{2}$

18. Find the area of an isosceles right triangle with perimeter $28 + \sqrt{392}$. (Hint: Use your answer from problem #17.)

Solution: $28 + \sqrt{392} = 28 + \sqrt{2 * 196} = 28 + 14\sqrt{2}$

s = 14; From #17: $2 * 14 + 14\sqrt{2} = 28 + 14\sqrt{2}$

$A = \frac{1}{2} l_1 l_2 = \frac{1}{2} * 14 * 14 = 98$

Answer: 98

19. Find the sum of the numerical value of the perimeter and the area of a square with a diagonal $\sqrt{120}$.

Solution: $s^2 + s^2 = 2s^2 = (\sqrt{120})^2 = 120$
$s^2 = 120 \div 2 = 60$
$s = \sqrt{60} = 2\sqrt{15}$
$P = 4s = 4 * 2\sqrt{15} = 8\sqrt{15}$
$A = s^2 = (2\sqrt{15})^2 = 60$

$P + A = 8\sqrt{15} + 60$

Answer: $8\sqrt{15} + 60$

20. Find the perimeter of the following pentagon. (Figure not drawn to scale)

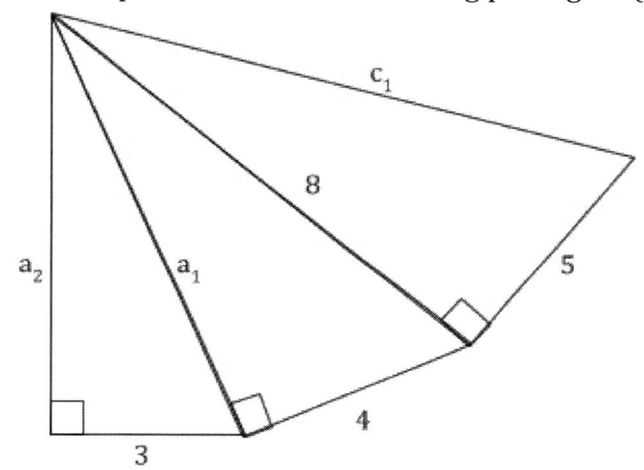

Solution: The right triangle on the right has legs 5 and 8: $c_1^2 = 5^2 + 8^2 = 25 + 64 = 89$
$c_1 = \sqrt{89}$

The right triangle in the middle has leg 4 and hypotenuse 8:
$a_1^2 = 8^2 - 4^2 = (8-4) * (8+4) = 4 * 12 = 48$
$a_1 = \sqrt{48} = 4\sqrt{3}$

The right triangle on the left has leg 3 and hypotenuse $4\sqrt{3}$:
$a_2^2 = (4\sqrt{3})^2 - 3^2 = 48 - 9 = 39$

$a_2 = \sqrt{39}$

Now, the perimeter can be calculated:
$P = \sqrt{39} + 3 + 4 + 5 + \sqrt{89} = 12 + \sqrt{39} + \sqrt{89}$

Answer: $12 + \sqrt{39} + \sqrt{89}$

About The Author

Jesse Doan started participating in math competitions in sixth grade. He earned perfect scores on the math section of the SAT and the PSAT. As an eighth grader, he scored a perfect 5 on the AP Calculus BC exam. Jesse is currently a high school senior from the island of Maui, Hawaii.

Below are some of Jesse's achievements:

- United States of America Junior Math Olympiad (USAJMO) Qualifier (2015)
- First Place Individual in Hawaii, American Invitational Mathematical Examination (AIME) (2015)
- First Place Individual in Hawaii, American Mathematics Competitions 12 (2016)
- First Place Individual in Hawaii, American Mathematics Competitions 10 (2014, 2015)
- Captain of the First Place Team, Hawaii State Math Bowl (2014, 2015, 2016)
- 48th Place Individual in the USA, MATHCOUNTS Nationals (2013)
- MATHCOUNTS Nationals Alumnus (2012, 2013)
- First Place Individual in Hawaii, MATHCOUNTS State Competition (2013)
- First Place Individual in Hawaii, American Mathematics Competitions 8 (2012)

Made in the USA
Columbia, SC
21 September 2020

21268126R00143